No School Left Behind

Based on the fieldwork carried out at two elementary schools, Merits School and Pioneer School, in northeastern China, the monograph details how local schools enacted the New Mathematics Curriculum Reform that was launched in early 2000.

The trajectory of the reform implementation at each school was plotted out. Both schools resorted to a long-standing quality control mechanism, i.e., teaching norms, to operationalize the reform ideas. The mechanism functioned by placing teachers under measurable supervision and evaluation aligned with the reform. The schools responded to the reform following school people's raising practical concerns, as well as the established school culture. Merits School arrived at a "two-faced strategy" to cope with the reform. Pioneer School managed to maintain a balance between promoting reform pedagogy and maintaining good test rankings. Both schools marginally involved parents in the implementation of the reform. This study suggests that to achieve success, reformers need to place equal emphasis on the transformation of teachers as well as local policymakers.

This book enriches the existing literature on the implementation of mathematics curriculum reform at the school level and brings insights into the schools' implementation decisions, which will appeal to policymakers, curriculum researchers and administrators.

Wei Gao, Ph.D., is an associate professor at the Institute of Higher Education at Beijing University of Technology, China. He obtained his doctorate from Syracuse University, United States of America. He has been an elementary teacher, UN Volunteer, special educator, and an engineer. His research is focused on STEM education and inclusion issues.

Xianwei Liu, Ph.D., is an associate professor at the Institute of Higher Education at Beijing University of Technology, China. Dr. Liu focuses on the area of STEM education, with a particular interest in students' capacity development in the STEM areas.

China Perspectives

The *China Perspectives* series focuses on translating and publishing works by leading Chinese scholars, writing about both global topics and China-related themes. It covers Humanities and Social Sciences, Education, Media and Psychology, as well as many interdisciplinary themes.

This is the first time any of these books have been published in English for international readers. The series aims to put forward a Chinese perspective, give insights into cutting-edge academic thinking in China, and inspire researchers globally.

To submit proposals, please contact the Taylor & Francis Publisher for the China Publishing Programme, Lian Sun (Lian.Sun@informa.com).

Titles in education currently include:

Innovation Competency Model
Shaping Faculty Academic Innovation Development in China's Higher Education
Jian Li

Research on Migrant Children's Educational Choices and Fiscal Policy
Hui Zhang

Barriers to Inclusive Education in Chinese Primary Schools
Culture, Policy and Practice
Qinyi Tan

Shifting Traditions of Childrearing in China
Narratives from Three Generations of Women
Xin Guo

No School Left Behind
Implementation of China's New Mathematics Curriculum Reform (2000–2020)
Wei Gao and Xianwei Liu

For more information, please visit www.routledge.com/China-Perspectives/book-series/CPH

No School Left Behind

Implementation of China's New Mathematics Curriculum Reform (2000–2020)

Wei Gao and Xianwei Liu

Routledge
Taylor & Francis Group
LONDON AND NEW YORK

The book is published with financial support from Humanities and Social Sciences Fund of the Ministry of Education of P. R. China (No. 17YJA880021)

First published in English 2021
by Routledge
2 Park Square, Milton Park, Abingdon, Oxon OX14 4RN

and by Routledge
605 Third Avenue, New York, NY 10158

Routledge is an imprint of the Taylor & Francis Group, an informa business

British Library Cataloguing in Publication Data
A catalogue record for this book is available from the British Library

Library of Congress Cataloging in Publication Data
A catalog record for this book has been requested

ISBN: 978-1-032-02335-9 (hbk)
ISBN: 978-1-032-02336-6 (pbk)
ISBN: 978-1-003-18294-8 (ebk)

Typeset in Times New Roman
by Newgen Publishing UK

To Tingyun, Xiangyun, and Yue

Contents

Figures

Tables

Preface

> "What is two divided by two thirds?," the education professor asked.
>
> "I know! I know!" Michelle raised her hand went up to the board, took the chalk, and began her computation.
>
> "2 divided by 2/3...multiply 3 here and here...cross 3...cross 2...Here you go. It is 3!"
>
> $$\frac{2}{\frac{2}{3}} = \frac{2 \times 3}{\frac{2}{3} \times 3} = \frac{2 \times 3}{2} = 3$$

Michelle was in her early 30s. She got her Master's degree from Teachers College before coming to Syracuse for her doctorate. It was a simple problem and a sixth grader in China would most likely give the answer right away. It took her almost 2 minutes.

In the United States (U.S.), it has been a popular saying that Chinese students outperform American students in mathematics, and mathematics education in the U.S. is at risk. Indeed, international assessments, such as the Programme for International Student Assessment (P.I.S.A.) of 2015 and 2018, show that students from China are ranked top. A number of factors are claimed to account for Chinese students' top mathematics performance (Wang & Lin, 2005). Stigler and Hiebert (1999) examined teachers' classroom instruction. They compared lessons taught by teachers from several countries and found that the American teachers spent a predominant amount of time on procedural matters instead of guiding students to reasoning and problem solving. They concluded that the U.S. teaching culture must be changed so as to improve students' mathematics achievement. Meanwhile, teachers in the U.S. are identified as particularly weak in mathematics knowledge. Liping Ma (1999) found that the majority of the Chinese teachers in her study demonstrated more in-depth conceptual understandings of mathematical content and had more knowledge of pedagogical content than the American teachers. It is believed that the Confucianism-oriented Chinese culture helps produce diligent pupils and dedicated teachers. China has a single national standards-based curriculum that prescribes uniform textbook contents across the country; even behaviors such as how a teacher opens his/her lesson is hinted to be attributable to the achievement gap between China and the U.S. (Fan et al., 2004). Such studies are informative, but, their limitation is that they treated mathematics education as compartmentalized activities and failed to see the whole picture of everyday mathematics teaching and learning in the school context.

Bearing this notion in mind, we were engaged in fieldwork at two elementary schools in a northeastern city in China and started to follow the implementation of the New Mathematics Curriculum Reform (referred to as "the reform" for short). The decade-long investigation revealed that the schools had a data-driven quality control mechanism to monitor, evaluate, and regulate teachers' activities throughout the whole teaching process. The data were systematically collected from reviews of lesson plans, observation of teachers' instruction and professional learning, examples of student work, and results of assessment. Those data provided ample evidence on teachers' performance and students' learning. Thus, teachers could receive timely feedback regarding their strengths and shortcomings, and make changes accordingly.

This mechanism is to ensure that no student, teacher, or school is left behind. The downside of the quality control mechanism could be that it was rigid, lacked flexibility, and even restrained individual creativity. Teachers might clone one another. Some teachers might be pressed by tight evaluation and become more concerned with their job security. They would assign an excessive amount of homework to drill students for good test scores at the cost of students' physical and mental health. P.I.S.A. data show students in China have the highest number of study hours in all participating countries. This might offer a potential explanation for the perplexing phenomenon that Chinese students performed much better than their American peers in solving procedural problems but were not as strong in open-ended problem-solving areas (Cai, 2004). The Reform meant to change this, but it has not got there yet.

1 Introduction

Since the early 1970s, constructivism has fundamentally transformed the understanding of mathematics and its learning and teaching (Steffe & Kieren, 1994). Mirroring the paradigm change in epistemology that perceives knowledge as interpretive and socially constructed (Van de Ven, 2007), traditional conception and practice vis-à-vis what mathematics is, how it should be learned, how it should be taught and so on, have been re-scrutinized. No longer being a set of universal truths, mathematics is understood to be socially constructed, life-related, and culturally important (Cobb, Yackel, & Wood, 1992). In light of constructivism, mathematics knowledge is not a fixed body of procedures and rules, but a growing and changing science of quantity and patterns, and the practice of mathematics involves experimentation, reasoning, and argumentation (National Research Council, 1990).

In school mathematics education, mathematics learning is no longer considered the mapping of the external world into the internal cognition of students via simple drill and rote memorization (Cobb, 1994). Instead, students are perceived as actively constructing the meaning of mathematical concepts via active interactions. Teachers' roles are expected to change from knowledge providers to facilitators, provocateurs, and questioners, who need to create an interactive environment, that is, a constructivist mathematics classroom, for students to model mathematically their lived worlds (Fosnot, 2005).

Different from traditional classrooms, constructivist classrooms should offer original curriculum materials, seek and value students' input, provide student-centered learning activities, and conduct authentic assessment of student learning (Brooks & Brooks, 1993). Ideally, constructivist mathematics classrooms should be like a workshop, a mini society, a community of learners, in which students are actively engaged in authentic mathematical activities, discourses, and reflections (Fosnot, 2005).

This constructivist vision for teaching and learning has impacted mathematics education to a great extent. One of the prominent influences can be found in the setting of mathematics curriculum standards in the U.S. Since the 1980s, education reform has been characterized by the development and adoption of curriculum standards at all levels of policymaking and governance (Fuhrman, 1999; Smith & O'Day, 1991). In such an era, constructivism continued to nourish school mathematics education. It culminated in the milestone document, the *Curriculum and Evaluation Standards for School Mathematics* (for short, the N.C.T.M. Standards) (National Council of Teachers of Mathematics (N.C.T.M.), 1989). The N.C.T.M. Standards (1989), grounded in constructivism, placed a greater emphasis on fostering students' conceptual understanding, reasoning, problem-solving, and communication. To this end, students need to actively interpret, organize, and construct the meaning of situations with mathematical modeling (Fosnot, 2005).

The N.C.T.M. Standards provided students and teachers a coherent and consistent roadmap. A common set of targets of knowledge and skills was specified for students to achieve, and the sequence of content areas was detailed for teachers to follow. In line with the N.C.T.M. Standards, many states in the U.S. developed mathematics curriculum standards that were constructivist-oriented and implemented standards-based reform curricula, in the hope of transforming instructional practices and enhancing academic achievement (Schoenfeld, 2004).

Since the first decade of this century, the Federal Government of the U.S. has been nationalizing curriculum standards benchmarked against challenging international ones. Arne Duncan, the then U.S. Secretary of Education, declared that "…the notion that we have 50 different goalposts is absolutely ridiculous. If we accomplish one thing in the coming years – it should be to eliminate the extreme variation in standards across America" (Duncan, 2009, para. 2). In 2010, the Common Core State Standards came into being (Common Core State Standards Initiative, 2010). Further efforts were made to develop a curriculum based on the Common Core State Standards, since innovating the curriculum has been the primary approach to inciting large-scale educational change for decades (Cohen & Hill, 2000; Darling-Hammond, 1990; Fullan & Pomfret, 1977; Stenhouse, 1975).

Even though standards and curriculum could be developed at the national level, it is local school districts and buildings, however, that determine the extent those instruments of reform succeed (Ogawa et al., 2003). China might be a potential site to examine how an innovative curriculum reform could be implemented nationwide and possibly enable instructional transformation. Interestingly enough, as the U.S. casts an appreciative eye over Chinese mathematics education (Asia Society, 2005; Stevenson & Stigler, 1994), China has grown increasingly dissatisfied with its own education and instead looks to the developed countries, particularly the U.S., as the role model for reform. Notably, following in the footsteps of the Western constructivist-minded reformers, China has embarked on a constructivist standards-based mathematics curriculum reform since the late 1990s.

Troubled basic education in China

Educational reform, according to Ginsburg and colleagues (Ginsburg et al., 1990), focuses on changing one or more of the following aspects of educational systems:

(1) Size or number of students, teachers, administrators, and buildings.
(2) Goals and objectives.
(3) Policy-making and the administrative/managerial system or power structure.
(4) Financing and budget-making processes.
(5) Level of funding.
(6) System organization: The types, statuses, and levels of as well as links and ages of transition between educational institutions.
(7) Curriculum: Content and organization of what is taught.
(8) Pedagogy: Social relations of teaching and learning.
(9) Selection, evaluation, and promotion criteria and procedures for students
(10) Selection, evaluation, and promotion criteria for educational workers (teachers, administrators, etc.).

(p.475)

In basic education in China, "reform" has been a buzzword since 1949. Seven rounds of nationwide educational reform was carried out until 1992 (Chen, 1999). On average, every 5 years there was a new round of curriculum reform. Even though some achievements were made, serious setbacks also occurred, for policies were constantly oscillating between whether education should be for economic development or for ideological integrity, for social equality or for efficiency, and/or for maximizing human capacity or for growing intellectuals (Tsang, 2000).

Previously, many reform efforts had been exerted in changing the curriculum and pedagogy. Yet, after many attempts, Chinese parents were disappointed to see that problems were persistent. Heavy examination pressure remained which haunted children, students were overloaded with testing in many subject areas, and teaching still relied on direct instruction with few interactions. One simple fact testifies to this observation. The Ministry of Education (M.O.E.) in China ordered to ease students' burdens (*jianfu*) 49 times from 1985 to 2000, however, the workload of students has only increased (Liu, 2007). People's dissatisfaction grew ever stronger with China's highly test-centric schooling.

In the late 1990s, two domestic and international factors paved the way for another round of educational reform in China. Firstly, China underwent rapid political, social, and economic transformation. As China changed from a primarily socialist planned economy to a market economy, growing consensus was reached with regard to the significance of education in advancing economic development, addressing social equity, and improving the qualities of human resources. Very possibly, there could be less resistance to and ambivalence about the direction of the reform. Secondly, in the meantime, China was expected to become a member of the World Trade Organization (W.T.O.) (China officially joined in late 2001) and face fierce competition with developed countries in the global market. The authorities stressed the imperativeness of reforming out-of-date basic education that suffocated the nation's originality and creativity, and impeded its development in science and technology.

Owing to China's unique relations with the West in history, the West not only serves as the source for its inspirations, but also stirs up feelings among the Chinese people. "Learning from the West", "chasing the West", and "beating the West"...these lofty slogans can always magnetize and energize Chinese society. Thus, the reference to globalization and competition with developed countries could effectively displace society's frustration with actual problems, and mobilize genuine support for the educational reform.

The Chinese central authorities began exploring the possibility of having another round of educational reform in 1996 (Huang, 2004). Under a new and soon to be widespread banner, *suzhi jiaoyu,* or "education for improving the qualities of students," the *Action Plan for Invigorating Education in the 21st Century* (hereinafter referred to as "the Action Plan") was formulated by M.O.E. and ratified by the State Council in 1999 (Guan & Meng, 2007). In order to greatly improve the education, *suzhi*, of the whole Chinese nation, the Action Plan spelled out the goals that should be realized in the realm of moral, intellectual, physical, and art education, early childhood education, and education for children with disabilities and ethnic minorities. Despite its vagueness, and danger of engendering and justifying social inequality, one key purpose of *suzhi jiaoyu* was to develop the competencies of children rather than focus on exam success, consequently leading to the revitalization of the nation. A subsequent policy document, the *Decision to Strengthen the Educational Reform and to Foster Full-scale Qualities-oriented Education* (hereinafter referred to as "the Decision") (M.O.E., 1999), marked the official launch of the eighth wave of comprehensive educational reform in China.

The Decision set forth goals in such aspects as educational equity, educational decentralization and privatization, promotion of lifelong learning, and internationalization and globalization (Guan & Meng, 2007). In all its sincerity, the Decision placed an emphasis on the need to substantially overhaul the national curriculum framework in response to the above issues. It directed that a new framework be remolded in line with advanced curriculum theories, and required that educational goals be reoriented and objectives redefined, curricular contents be updated, accessible, related to real life, and promote higher-order thinking, and instruction be explorative, heuristic, and child-centered. The Decision stated that the curriculum should consist of three tiers of management at the national, local, and school levels.

Development and implementation of the National Mathematics Curriculum Standards

The preparation stage (1999–2001)

The first move was to replace the old teaching outlines with curriculum standards. In China, the teaching outlines were documents that outlined the sequence, scope and depth of knowledge of each subject area at each grade level, which teachers, students, and textbook writers should stick to. New curriculum standards were to set minimum expectations for teaching and learning. The task of designing new curriculum standards for 18 subject areas from Grades 1 to 9 started in March 1999. M.O.E. called up an expert group that consisted of 15 university professors, several Grade 1–9 mathematics teachers, and a number of mathematicians, publishers, and teacher trainers. The group explored a wide range of issues. For instance, they identified the problems in the current curriculum, researched the reform practices of developed countries, and envisioned mathematics for the next century. After a dozen revisions, the group produced the trial version of the National Mathematics Curriculum Standards for Grades 1 to 9 (for short, the N.M.C. Standards) in February 2000 and sent out 40,000 copies for feedback in March of that year (China Education Network, 2005). Also, the N.M.C. Standards (Trial) was officially published in June 2001 (M.O.E., 2001). Between the creation of the trial version and its official publication, textbook series with newly added content were piloted in more than 40 schools in Beijing, Guangdong, and Xinjiang. Different from the previous "one outline, one set of textbooks" policy, the central authorities allowed having multiple textbook series written in accordance with the Standards. Altogether, eight sets of textbooks were adopted for the elementary level, and six for the secondary level.

The pilot stage (2001–2004)

In September 2001, the beginning of the new school term, the N.M.C. Standards were tried out in 42 experimental counties selected by M.O.E. Before that, backbone teachers ("mainstay" teachers, the selected core teachers with potential), principals, and teacher trainers selected from these regions were brought to the national training centers for multiform professional development activities by a number of trainers. The developers of the N.M.C. Standards directly administered over 40-hours of training for those backbone teachers who would themselves become reform demonstrators back in their own regions (Yuan, 2005). One year later, about 520 province-level pilot programs, or 18% of all the counties in China, were established and started using the N.M.C. Provincial authorities were in charge of all reform-related activities under the supervision of M.O.E. Until September 2003, more than 1,400 counties, nearly 50% of all counties, participated in the reform.

Full implementation (2005–2010)

In the years that followed, the reform was carried out in a sweeping manner. Meanwhile, different opinions regarding the N.M.C. were voiced sporadically and some mathematicians even appealed to the central government, calling for an immediate termination of the N.M.C. (Jiang, 2005). Yet, there was no sign that the N.M.C. would be paused; on the contrary, almost 90% of the counties in China started using the N.M.C. in fall 2005. A modified version of the N.M.C. Standards was published 2 years later (M.O.E., 2007). As a periodic summary, M.O.E. held a national meeting in 2009 to reflect upon the curriculum reform. The government called for taking a further step in the curriculum reform in 2010.

The revision of the core competences (2011–the present)

The N.M.C. Standards were revised and officially published in 2011. The revised draft has some modifications in the basic concepts, curriculum objectives, curriculum content, and implementation suggestions. It was adopted from then on. In April 2014, M.O.E. (2014) echoed the central government's call for comprehensively deepening curriculum reform to fulfill the fundamental task of cultivating moral and competent persons through education. The notion of core competencies was popularized. In mathematics, students were expected to develop core competencies such as mathematical abstraction, logical reasoning, mathematical modeling, mathematical operation, intuitive imagination, and data analysis.

The guiding principles of the National Mathematics Curriculum Standards (Trial)

The N.M.C. Standards (Trial) treated Grade 1 to Grade 9 as an integral process and consisted of three bands: Grades 1–3, Grades 4–6, and Grades 7–9. The N.M.C. Standards defined *basic knowledge and skills*, *mathematical thinking*, *problem-solving*, and *affection and attitudes (e.g. curiosity and confidence)* as four dimensions of teaching and learning objectives. It used such verbs as *know*, *understand*, *master*, and *flexibly apply* to characterize the degree of knowledge and skills being learned; it used *feel*, *experience*, and *explore* to describe the processes of mathematical thinking, problem-solving, and affection and attitudes. Students are expected to acquire important mathematics (including mathematical facts and experiences) necessary for their future social life and further development, as well as basic mathematical thinking and necessary skills.

The document contained four Content Standards representing goals in content areas of number and algebra, space and figures, statistics and probability, and practicum and comprehensive application, to replace the traditional structure of number and operation, quantity and measurement, basics of geometry, basics of algebra, ratio and proportion, applied problems, and basics of statistics. The N.M.C. Standards applied to all bands with different emphases. Some content areas in the traditional mathematics curriculum were reduced, such as applied problems (due to being overcomplicated and fabricated), operations of multidigit (more than three) integers, decimals, and fractions. Some content areas were added extensively. Grade 1 teachers started teaching shapes and figures, geometry transformation (reflection, translation, and revolution), and data collection and analysis.

Textbooks were redesigned accordingly. One of the most used sets was published by the People's Education Press. Lu and Wang (2004) summarized that this set of textbooks demonstrated several distinctive features. The central tenets were: (1) Attending to the process of learning and children's attitudes, (2) enabling knowledge construction based on children's prior experience and acquired knowledge, and (3) emphasizing problem-solving and inquiry. The textbooks incorporated broader mathematical knowledge like basic statistics and probability. The presentation of content was more vivid and light with many cartoons and pictures. More life-related activities and practical problems were designed, for example, children were given the chance to collect and analyze data about their favorite food, extracurricular activities, or television programs.

The N.M.C. Standards (M.O.E., 2001) set forth six guiding principles on areas such as curriculum, ontology of mathematics and utilities, teaching, content and learning, assessment, and technology. In the following, we discuss these general statements one by one. It states:

(1) The school mathematics curriculum should highlight its fundamentality, universality, and developmentality so as to make mathematics education relevant to all students. It should fulfill these goals: (i) Everybody learns mathematics worth learning, (ii) everybody can acquire the mathematics necessary for him or her, and (iii) different people can achieve differently in mathematics. Characterized as fundamental, accessible, and developmental, for the first time, the N.M.C. Standards specify that mathematics should be for all (not for the minority elite). It also consciously differentiates mathematics as being worthy from being unworthy, and being necessary from being not needed, assuming that different students will inevitably lead to different learning outcomes. Yet, what is deemed as *worthy* and *necessary* mathematics might be a matter of arbitrary judgment.

(2) Mathematics is indispensable in our daily life, work, and study. It helps us process data, compute, reason, and prove. Mathematical models can describe effectively natural and social phenomena. Mathematics provides a language, thinking, and methods for other sciences, and constitutes the foundation of all major technological advances. It plays a special role in enhancing human reasoning, abstraction, imagination, and creative capacities. Mathematics is a part of human cultures, and its contents, thinking, methods, and language are an important part of modern civilization.

Mathematics is accorded fundamental roles and applications by the Standards. The nature of mathematics, or the ontological question of mathematics, is vaguely touched upon. It explicitly claims that "mathematics is a culture," suggesting that mathematics is a social product instead of being something there a priori. This view implicitly echoes the humanistic or social constructivist positions in that mathematics is understood as a sociocultural phenomenon (Bishop, 1994; Prediger, 2004).

In terms of mathematics content and learning, it says:

(3) The contents of school mathematics should be realistic, meaningful, and challenging. They should help children to engage in such mathematical activities as observing, experimenting, guessing, verifying, reasoning, and communicating. Contents should be presented in different ways to satisfy the diverse learning needs of students. Effective learning activities should not merely rely on imitation and rote memorization. Hands-on activities, self-initiated exploration, and cooperative learning are important approaches to learning. Owing to variations in students' cultural environments, family backgrounds, and disposition, mathematics learning should be an animated, lovely, proactive, and unique process [for everyone].

This principle contains rich and key information. Not only does it stipulate what mathematics knowledge be included in school mathematics education and what effects it should result in, but also it portrays the process of mathematics learning. Observing, experimenting, hypothesizing, and verifying are emphasized as important activities to develop mathematical thinking and reasoning (informal and deductive). Active hands-on approaches, inquiries, and cooperation between students are emphasized. In addition, communication is also mentioned (though not particularly underscored) as one means of learning. Most importantly, the N.M.C. Standards view mathematics learning as a *process* rather than merely a result, which should take into account students' socioeconomic and cultural uniqueness. Dynamic and explorative, learning mathematics is no longer regarded as an acultural, purely intellectual performance.

(4) Mathematics instruction must be based on the levels of students' cognitive development as well as prior knowledge and experience. Teachers should stimulate children's motivation for learning and provide them ample opportunities to participate in mathematical activities. Teachers should help children truly understand and grasp basic knowledge and basic skills, mathematical thinking and methods, and acquire broad mathematical experiences through self-exploration and cooperative learning. Students are the masters of their own mathematics learning, and teachers are the organizers, guides, and collaborators of mathematics learning.

Instruction being developmentally appropriate is required. The above proclamation echoes the Piagetian theory that conceptualizes children's cognitive development into four stages. Teachers also bear the responsibility of creating educational situations for inquiry and exploration so that students can master: Basic knowledge, basic skills, basic mathematical thinking and methods, and basic experiences.

The roles of teachers and students are clearly defined and distinguished. The N.M.C. Standards maintain that students should take back the ownership of their learning with facilitation and guidance from teachers, and teachers should no longer act as the knowledge transmitters or even nannies to spoon-feed students. However philosophically thought-provoking and well-meaning, it raises new issues. As discovered in reality, this distinction causes many teachers to freeze and not dare to play active roles in teaching. They seem to be marginalized instead.

(5) The main purpose of assessment is to understand globally students' mathematics learning through the process, to motivate children for learning, and to improve teachers' instruction. An assessment system with multiple goals and means should be established, which not only cares about students' learning outcomes, but more about the process; not only about their mathematical proficiency, but about their affections and attitudes exhibited during mathematical activities. Teachers need to help students know themselves and build up their confidence.

Of course, this principle broadens the current notion of assessment characterized by high-stake tests that alienate students from authentic learning (Agrey, 2004). Yet, this statement does not offer any guidance or suggestions regarding what forms of assessment are plausible to assess the learning process. Textbook designers and teachers are left alone to interpret this information on their own.

Lastly, the principle of integrating information technologies in mathematics education is proposed:

(6) The development of modern information technologies has great impact upon the values, purposes, contents, and pedagogy of mathematics education. The design and implementation of the mathematics curriculum should attend to applications of modern information technologies, and fully consider how calculators and computers have influenced on the contents of and approaches to mathematics learning, [so as to] develop and provide richer learning resources to students. I.T. should become a powerful tool for mathematics learning and problem-solving. Teachers should endeavor to change students' learning habits so that they are willing to invest in and more enthusiastic about realistic and exploratory mathematical activities.

Controversies over the National Mathematics Curriculum Standards

The N.M.C. Standards encountered much of the resistance from the mathematician community. Jiang Boju (2005), a fellow of the Chinese Academy of Science and a mathematics professor at Peking University, was a symbolic figure who opposed the N.M.C. He warned that the N.M.C. would bring disastrous consequences to mathematics education in China because it would erode the solid mathematics foundations of students in the country. There were two fatal issues with the N.M.C., according to him. Firstly, the N.M.C. was an attempt to "get back to the drawing board" (*lingqi luzao*, or, setting up a new cooking station), which disregarded the traditional, time-honored, domain-specific presentation of mathematical knowledge. Joining in other critics, Jiang criticized the N.M.C. for being loose and chaotic, because of its reorganization of mathematics into four parallel areas. By teaching all four content areas at every grade level over the nine grade levels and treating none to an in-depth degree, he posited that the N.M.C. underestimated children's intellectual potential and complicated the work of teachers: "In the name of spirally growing…to learn something deeper, you have to wait one more year for the next circle!" (Jiang, 2005). As a consequence, the reform would lead to nowhere and make "teachers too confused to teach and students too dumbed to learn" (Jiang, 2005).

Secondly, Jiang strongly voiced his concern that the N.M.C. watered down the prime contents of mathematics due to the reformers' cries for "decreasing the level of difficulty," "connecting mathematics to students' real life," and "mingling life and mathematics together" (Jiang, 2005). Take the secondary mathematics as an example, he argued, the N.M.C. replaced the commonly used concept of *plane geometry* with the less mathematically precise term, *figures*, and it also lowered the requirements on deductive reasoning and proof. Such moves might adversely hurdle the development of children's logical thinking, since "mathematical proof is the soul of mathematics" (Jiang, 2005, para. 6).

Elevating his argument to another level, Jiang cautioned that school mathematics should have a fundamental status in basic education, and in particular it shouldered the mission of enlightening the Chinese people in terms of rational and scientific civilization for the invigoration of the nation. However, the N.M.C. seemed unfit for this purpose, since it discarded the precious tradition of China's school mathematics education. A system like this might hardly teach children to learn how to analyze problems and develop their logical reasoning, never mind nurture their innovation abilities.

The reformers and supporters discredited such charges. He (2006) argued that there was an urgent need to rebuild a new mathematics curriculum. Many attempts had been made to patch up the previous complicated, difficult, irrelevant, and out-of-date curriculum, but students still did not like mathematics and China was still far inferior to first-class countries in mathematics research. It was necessary to construct a totally new cooking station, he concluded. In accordance with others (Sun, 2005), He (2006) continued to argue that by synthesizing different content areas, the new curricular structure was superior to the traditional domain-specific system, and reflected the relationships between mathematics and nature, mathematics and society, and mathematics and the sciences. Moreover, they emphasized that the N.M.C. drew extensively upon the reform experiences of developed countries, suggesting that it was proven to be more advanced.

In response to the issue of whether the N.M.C. lowered the level of difficulty, the reformers defended their positions on different occasions. He (2008) posited that in such an era of information, there was more important information waiting for children to process than pure mathematical knowledge and it was not necessary for everyone to learn axiomatic mathematics. The key was to develop and preserve children's ability to identify, formulate, and solve problems that were not given in formal mathematics textbooks. That being said, the N.M.C. has put more emphasis on preparing children to "try, observe, estimate, compare, synthesize, imagine, and hypothesize" (M.O.E., 2001, p.4).

Jiang's second criticism seemed to concern one of the reform's fundamental principles – pursuing educational equity or *making mathematics for all*. Would lowering the difficulty of the N.M.C be conducive enough to make mathematics accessible to all? Or would it be destructive instead?

Sun (2005) espoused the math for all principle by drawing a distinction between floor math (for the masses) and ceiling math (for the elites). He stated in length:

> To my understanding, figuratively, if math education is one room, math for the masses is like the floor of the room. It is 'floor math'. Everyone should start learning to stand on the floor firmly and stably. The ceiling is math for the elites. In between, there exists a vast space to pursue. Only standing steadily on the floor, propelled by strong motivation, having potentials and talents, can children have the possibility to jump up and touch the ceiling…the masses potentially are the bases where mathematicians and scientists emerge.
>
> (para. 4)

Ginsburg et al. (1990) stress that educational reform viewed from one perspective might be improving the effectiveness, efficiency, and relevance of the educational system, but in fact may "increase or at least reinforce inequalities in education and society" (p.475). This might be the case for the N.M.C. Standards. Theoretically speaking, Sun's defense for "math for all" stands firm. Nonetheless, one important cultural factor is overlooked, in that, what the majority of Chinese parents want and propel their children to achieve is the ceiling. The action of decreasing the difficulty of the N.M.C. might cause an unintended consequence, widening the inequality between students. Complaints are often heard that nowadays teachers teach less concrete mathematics during school hours. Therefore, a large number of students have to seek after-school paid tutoring to avoid being left behind in ever competitive examinations. Different from the fairly equal competition for opportunities between students before the reform, which were up to individual diligence and intelligence, the parental or family factor begins to play an increasingly significant role. Since poorer families generally lack the social and cultural capitals and possess fewer resources to dispose, they can provide little intellectual support to their children, and they cannot afford extra educational expenses to purchase services beyond schools. In a sense, who can touch the ceiling math becomes a competition between the parents and families.

One more issue that confronts the reformers is that the theories and ideas guiding the N.M.C. mostly originate from Western societies (Cha, 2007). The century-long mentality of "learning from the West in order to beat the West" unproblematizes the behaviors of extensively borrowing paradigms of thinking and practices from the West. In fact, Western scholars have pointed to the dominant influence of developed nations in producing and diffusing ideas and practices that cause the less developed countries to borrow and converge with the West (Inkeles & Sirowy, as cited in Ginsburg et al., 1990). The lack of independent thinking and careful consideration of Chinese unique history, culture, and traditions in formulating educational polices raised the concern that China was in the danger of cultural and educational neocolonialism (Liu, 2006). For instance, the N.M.C. Standards (Trial) had many similarities with the N.C.T.M. Standards (2000) in the U.S. Both were premised on the tenet of constructivism. Both set forth similar principles. Both broke the traditional domain-specific structure of mathematical content areas. Both abandoned the traditional order of content provision and all content areas were offered at all grade levels. Both advocated for student-centered, inquiry-oriented, cooperative learning. Both stressed the importance of relating mathematics to children's real life. And both placed an emphasis on the mastery of basic knowledge, facts, and skills, but paid more attention to conceptual understanding.

The N.M.C.'s departure from the past gave rise to many debates. Despite the concerns voiced, the reform marches forward. The N.M.C. was in place in most schools until 2005. There were many reasons to believe that the Reform would succeed in fulfilling *suzhi jiaoyu*, since the net enrollment rate of higher education institutes, the once believed bottleneck of quality education, has surged to 49% in 1999 from 6.1% in 1979 (China National Knowledge Infrastructure, 2019). There were also many reasons to believe that this reform would unquestionably flunk, since it once again tried to repair only the curricular system . Controversies did not go away overnight. The opponents' attempts to boycott the new mathematics curriculum lingered on. And the reformers were determined to refute all attacks against the N.M.C. from the conservatives; they would unyieldingly carry the reform forward (Zhong, 2006).

The study

The purpose of the study

As was the case of the U.S. in the early phase of constructivist mathematics curriculum reform, debates between advocates and critics have rarely been informed by empirical research, and claims on both sides are largely based on hypotheses and speculation (Porter, 1994). In the "math wars" in China those who made their viewpoints heard in the media are rarely those who have been affected most by the N.M.C. reform. Grounded in structural symbolic interactionism (Stryker, 2008), and informed by curriculum implementation theories (e.g. Cohen & Ball, 1990; Spillane, 1996; Stenhouse, 1975), we employed the case study methodology (Yin, 2003) to explore the experiences of Chinese local elementary schools in response to the N.M.C. reform. This exploration was guided by the overarching research question: How do schools at the local level experience and act upon the N.M.C. reform? To pursue this inquiry, we conducted fieldwork at two elementary schools in one northeastern city in China. Bearing in mind the big picture, the book takes a longitudinal view of the reform in mathematics education in particular, and evaluated the reform based on the two specific cases.

It's worth noting here that China's New Curriculum Reform is not drawing the final curtain yet. It is a consistent endeavor of reformers and school people taking place over more than 20 years, with groundbreaking achievements as well as setbacks. Thus, the 2001 Trial and 2011 Revision should not be treated as products of two separate reforms, but the refined product at two different stages of the reform. The 2011 Revision of the N.M.C. Standards took on the legacy of the 2001 Trial. It did not overturn the core ideas, instead it was more of a refinement (and compromise to a certain degree) of the 2001 one.

Organization of the book

Chapter 1 introduces the international and domestic context of the Chinese mathematics curriculum reform and presents the rationale for the study. Chapter 2 is devoted to a detailed review of literature on curriculum innovation theories and practices, and teacher professional development models. These two bodies of literature framed the study, since the ultimate goal of curricular and instructional change can only be made possible through the professional development of teachers. Chapter 3 describes the design of the study and addresses methodological and cross-cultural ethical issues that the study encountered.

The remaining chapters report on the two cases and the cross-case analysis. Specifically, Chapters 4–9, respectively, present the cases of Merits School and of Pioneer School. The reform trajectory at each site is traced, the dynamics of the reaction of administrators and teachers to the curriculum reform are revealed, professional development activities for teachers are depicted, and the changes made and issues incurred are highlighted. Chapter 10 summarizes the findings of the two cases, and Chapter 11 presents a general discussion. Chapter 12 is the concluding chapter in which the implications and limitations of the study are addressed.

2 Curriculum reform as leverage of change

Reforming the curriculum to leverage instructional change and improve student achievement is not a new strategy – it has been a continuously engaging and intriguing field over nearly half a century (Cobb & Jackson, 2011; Cohen & Hill, 2000; Darling-Hammond, 1990; Fullan & Pomfret, 1977; Ogawa et al., 2003; Spillane, 1999; Stenhouse, 1975). The implementation of curriculum reform is a highly complex process. It entails "change in (1) subject matter or materials, (2) organizational structure, (3) role/behavior, (4) knowledge and understanding, and (5) value internalization" (Fullan & Pomfret, 1977, p.361). Throughout the process, a variety of stakeholders are involved, from higher-level policymakers, curriculum designers, commercial publishers, school district administrators, building principals, to classroom teachers, students and parents (Darling-Hammond, 1990).

A shift in the study of implementation of curriculum change is seen around the 1980s from a functionalist "input–output" perspective to a sociocultural and interactionist perspective. Before, most research in this field focused "either on characteristics of the policy or characteristics of the individual as determinants of implementation success or failure" (Drake, 2002, p.312). The interactions between curriculum reform policies and local implementers has been missed in functionalist studies. As Darling-Hammond (1990) posits:

> The 'input–output' mode of policy analysis did not examine how policy recipients at different levels (school board members, central office administrators, school principals, teachers, parents, students) viewed or experienced the delivered wisdom of legislators and bureaucrats, how they sought to understand and incorporate new structures into their work, or even whether the policy whose effects were to be measured had been implemented at all. The approach failed to ask what other events and conditions in the environment might support or undermine the intentions of the policy or might have independent effects on the outcomes of interest.
>
> (p.340)

Earlier attempts (Darling-Hammond, 1990) were made to understand the local meanings of reform mandates by taking into account the views and experiences of local implementers. More recently, researchers have paid increasing attention to the intriguing relationships between the implementation of reform initiatives and sensemaking of local implementers, for instance, classroom teachers (Cohen & Ball, 1990; Spillane, 1999). Interpretations that local implementers make have been found to mediate policy implementation; however, such interpretations might be framed and constrained by teachers' and administrators' understandings of who they are (that is, their role identities), prior experiences, and larger sociocultural and structural forces (Drake, 2002). This chapter reviews a number of sociocultural and interactionist studies on curriculum change, specifically, those concerning mathematics curriculum reform. These studies supply the necessary sources of thinking and concepts and theories to this study.

Teachers as key agents of change

Teachers' prior belief and practices mediating instructional change

Much research on curriculum reform has initially focused on teachers, since teachers are considered the key arbiters of instructional content and practice (Cohen & Hill, 2000). Teachers' instructional change is found to be mediated by teachers' existing attitudes, beliefs, and traditional practices. In other words, teachers "translate their knowledge of mathematics and pedagogy into practice through the filter of their beliefs" (Manouchehri, 1997, p.198). Teachers' belief systems refer to personal theories about the nature of teaching and learning of mathematics, which influence teachers' curriculum decision-making, and affect their classroom practices in a highly complex and dialectical fashion (Pajares, 1992). In other words, beliefs and practices reciprocally influence each other.

A number of studies have been done by Ball and colleagues (Ball, 1990; Cohen & Ball, 1990) to probe the implementation of the *Mathematics Framework for California Public Schools: Kindergarten through Grade 12*. A special issue of *Educational Evaluation and Policy Analysis* (1990) was devoted to examining individual classroom teachers' responses to the Framework. Five case teachers were reported on. It was found that the change of instruction by those teachers was filtered through their established practices and beliefs about mathematics teaching and learning. They demonstrated a mixture of old and new mathematics instruction. For example, one teacher thought that she had made significant change in line with reform ideas, while concrete materials like manipulatives were used in a traditional way to simply help students memorize rules and procedures (Ball, 1990). Some teachers adapted the curriculum to their own approach to teaching as well as to support their own view of mathematics. Some teachers did not offer students the opportunities to apprehend the N.M.C. or to construct mathematics as the Framework advocates. Others omitted new topics, such as probability, because of their belief that students should learn the basics first. Cohen and Ball (1990) concluded that teachers do not simply assimilate new curricula or change their practices as a passive response to the imposed reform. Instead, teachers make individual sense of and enact new policies in light of their inherent knowledge, beliefs, and practices. Similarly, Fuhrman (1999) noted that in implementing the N.M.C. standards, teachers' perceptions of their own capacity and their students significantly influenced how they responded to the standards and what reform ideas they took on board to improve their instruction. Teachers out of the "practicality ethic" might try only those new ideas that directly contribute to their present classroom activities (Doyle & Ponder, 1977–1978).

Teaching experiences affecting reform implementation

Accumulating evidence (Cohen & Ball, 1990; Fuhrman, 1999) has shown that instruction is multidimensional not monolithic. Some dimensions of instruction might be revised more readily and rapidly than others. As such, the pattern of instructional change might be different for teachers with varied teaching experiences.

Spillane and Zeuli (1999) observed 25 classroom teachers in Michigan, who reported more reform-oriented practice, to identify patterns of change in their instructional practices. The study employed conceptual and procedural mathematical knowledge, academic task, and discourse patterns as the conceptual framework to explore the progress of practice change. Out of the 25 teachers, only four were observed to have demonstrated practices reflecting the genuine reform ideas, and dismantled behavioral regularities (that is, teacher as teller, student as listener working mostly as a whole class or on their own) and epistemological regularities (that is, a focus on procedural mathematical content and doing mathematics as mostly computation) of instruction. Those four teachers set up mathematical tasks and renewed discourse norms that helped students develop conceptual knowledge of mathematics. Another ten teachers used somewhat concept-oriented tasks but fundamentally limited themselves to procedural discourse that aimed at *the* right answer. The tasks and discourse norms for the other 11 teachers, or 44%, "were firmly grounded in procedural mathematical knowledge and computational skills" (p.19).

Spillane and Zeuli's study (1999) suggests that teachers might undergo different stages of change. At the same time, it underscores the strenuous complexity of inciting change in instructional practices. However, readers might still question why almost half of teachers did not enact any of the reform ideals.

One possible explanation might be the number of years of teaching experience. Drake (2002) used career stages, defined by years of formal teaching experience, as the framework to identity patterns in teachers' responses to mathematics education reform. He revisited the data in Cohen and Hill (2000), and examined teachers' disposition to and understanding of the reform, preparation to teach in reform-minded ways, as well as the use of traditional and reformed teaching practice according to that career stage, professional development time, and the socioeconomic status of students. Early career teachers (1–3 years' experience) were found to be most favorable to mathematics education reform, but they had the least knowledge of reform ideas and felt the least prepared to teach in reform pedagogy. The levels of reform practices between teachers at different career stages were not distinguishable, though late career teachers had the lowest perceived reform practices, and mid-career teachers reported the highest levels of reform practices which, the author conjectured, might be attributable to their desire to change in order to avoid boredom and stagnancy in practices.

To understand how and why teachers at a certain career stage might make sense of and implement reform differently, the author interviewed six teachers, two from each of the three career stages: Early (1–3 years), early/mid-career (3–5 years), and mid-career (5–10 years). From the limited number of participants, the author noted that early career teachers were more prone to focus on procedural and basic aspects of practice because of the lack of competence or time to address conceptual core mathematical ideas of the curriculum. However, early/mid-career teachers viewed reform more comprehensively and purposefully managed to integrate the principles of reform into their teaching. Mid-career teachers integrated conceptual and procedural aspects of reform into their existing traditional practices in a piecemeal fashion. It was also observed that mid-career teachers had less reliance on authority or external sources of information, such as tests, standards, or written curricula, partly because they had firmly established their own beliefs and practices. The finding again corroborates what previous studies have repeatedly said: Prior beliefs matter.

Drake's study (2002) suggests that there seems to be a general pattern for teachers with different years of experience in responding to educational reforms. Reform designers could provide interventions to teachers at certain stages. However, considering that the Framework in California had been in place for 9 years when the data were collected, it means most early/mid- to mid-career teachers started their teaching career with or after the new standards being implemented. In contrast, mid-career teachers began to be exposed to the reform ideas in the middle of their career. It might be worth exploring how being exposed to the reform at different stages might impact teachers' accepting new ideas.

Teachers' zones of enactment

In spite of the knowledge learned about teachers' beliefs, prior practice and experiences mediating their enactment of curriculum change, it is still unclear why some teachers appear to change more drastically than others. Spillane (1999) proposed the model of teachers' *zones of enactment* to address this matter. Enactment zones, he argues, are the space "where teachers' personal capacity, will, and prior practice interact with incentives and learning opportunities mobilized by the policy, professional, public and private sectors, as well as incentives and disincentives for change teachers perceive in pupils' responses to instruction" (p.144).

Spillane used this model to revisit the teachers in one of his early studies (Spillane & Zeuli, 1999). The zones of enactment of 25 teachers were examined. Among the teachers studied, only three teachers had significantly changed the core of their practices. A deeper investigation revealed that ongoing collegial collaborations within and outside the classroom, school, and district heavily accounted for the change in the core practices of those three teachers. In other words, professional learning opportunities within teachers' zones of enactment were more favorable to facilitate teachers' change. Spillane (1999) claimed that "the extent to which teachers revise the core of their practice depends on their enactment zones" (p.144).

In summary, teachers' personal theories of teaching and learning are closely linked to instructional approaches they adopt in class. Reflecting varied degrees to which they aligned with the reform spirit, teachers show different patterns of change. This phenomenon might be attributable to teachers' experiences of teaching. Without any doubt, the whole microecology in which teachers encounter reform policies is not devoid of the influence of their own experience.

District and school administrators as policymakers

Successful implementation of instructional reforms and revision of teachers' core practice depend in considerable part on local policy environments (Ogawa et al., 2003; Spillane, 1996; Spillane & Thompson, 1997). District and school administrators do not passively implement policies from the higher authorities. On the contrary, they actively reinterpret those policies and make their own ones.

Using rational and institutional perspectives, Ogawa and colleagues (Ogawa et al., 2003) investigated the adoption, development, and implementation of a standards-based mathematics curriculum and criterion-referenced assessment at one school district in California. The researchers interviewed district administrators, school principals, and reform teachers, and collected documents and artifacts. They identified that the standards-based curriculum was implemented more symbolically than rationally by district administrators and echoed by school administrators. Such symbolic behavior was meant for the district to gain legitimacy so that it could survive in the changing policy environment. For instance, the district adopted curriculum standards that were less rigorous than the state's standards in an attempt to be true to local situation, which nonetheless might perpetuate students' lower performance. In developing the standards, classroom teachers were marginally involved, not to capitalize on their expertise but to gain greater buy-in from them.

Throughout the implementation, the district did not, and could not, provide a clear, cohesive instructional technology, which rendered professional development, model lessons, and oversight of school administrators superficial rather than substantive. Thus, teachers taught to the standards based on varied understandings. The authors of the study concluded that ultimately the decisions and actions of local districts and schools have the most influence on how reform ideas affect teachers' practice[1]. Their finding is reminiscent of the postulate of Price, Ball, and Luks (1995) that if high-level administrators did not treat the reform sincerely, its implementation would be hard to sustain and teachers would very likely retain their traditional values and practices, but, why high-level administrators decide to make such decisions needs further exploration.

School districts matter because they actively interpret policies stipulated by higher-level agencies, and make their own policies accordingly. That is, districts perform not solely as policy implementers but as policymakers too (Spillane, 1996). Spillane (1996) examined the implementation of a state reading policy in two school districts by drawing on three sources of data: Interview, document analysis, and observation. Both districts were observed not as passively conforming to state policies, but as taking an active stance in interpreting the state policies and making their own versions. For example, one district put forth more policies in line with the state policy, and expanded the opportunities for teachers to learn reformed instructional approaches. However, the other district "buffered out many of state policymakers' instructional reform ideas" by decoupling professional development resources (e.g. textbooks) from state reform ideas (p.82). One essential message derived from this study is that local school districts could undermine or expand policies through their local policymaking efforts. Policymaking needs to conscientiously take local circumstances into consideration in order to streamline instructional guidance at multiple policymaking levels.

Active leadership of local authorities can be instrumental to facilitate district-wide instructional change. Spillane and Thompson (1997) studied nine school districts in Michigan that took on instructional reform in mathematics and science education. One rural school district distinguished itself from the others. Even though the district was resource-limited, the local education agency – that is, district administrators, principals, and teacher leaders – was proactive in promoting the new instructional technology. They sought external experts from outside the district, forged trust among teachers, provided consistent professional learning opportunities, freed up time for teachers, and used new curriculum materials as the common learning agent. Consequently, both district and school administrators, and teachers could learn interactively the substantive reform ideas and foster mutual learning within the district.

Policies do not implement themselves in an automatic and straightforward fashion; instead, specific human beings are behind the curtain to operationalize them. Through their individual sensemaking and manipulation of the same piece of well-intended policy, district and school leaders are not only able to promote but they are also able to undermine classroom teachers' professional interactions and learning in more subtle, but seminal, ways. Coburn and Russell (2017) attempted to disentangle the intricate dynamics among district professional development initiatives, district and school leadership, and teachers' change. They resorted to a social capital framework, and scrutinized teachers' social networks in order to determine how district and school leaders might have influenced teachers' interactions within their social networks. Four dimensions of teachers' social networks are deemed sources of social capital: Structure of ties, trust, access to expertise, and content of interaction. The study looked at two school districts, Greene School District and Region Z, each with four case study schools. Both districts had recently embarked on a district-wide scale-up of a new standards-based elementary mathematics curriculum. Participants mainly consisted of six focal teachers, eight principals, and 13 math coaches (nine for Greene School District and four for Region Z). The study yielded three major findings. First, both districts assigned coaches available to teachers at every school, which increased teachers' tie span. But, teachers in different districts resulted in having different access to expertise. Leaders at Z District selected coaches in such a haphazard way that most coaches had no background in the curriculum and little prior professional development in mathematics teaching and learning, while all the coaches at Greene had either moderate or high expertise thanks to clear selection criteria. Second, Region Z and Greene had different policies regarding what coaches should do at the school site, which affected the strength of teachers' ties with coaches and between colleagues. Mathematics leaders in Region Z did not clearly articulate coaching tasks, while Greene made it clear to all school leaders and coaches what mathematical tasks classroom coaching with teachers entailed. Third, the routines of interaction that districts enforced influenced the depth of interaction in focal teachers' social networks. District leaders in Greene engaged coaches in task analysis, analyzing student strategies, and structured reflection, while Region Z emphasized explanation, doing mathematics problems to learn how to do them, and mapping activities. These results suggest that having policies in place is not enough; rather, those who design and carry out the initiatives in the first place might determine the quality and outcomes of experiences of teachers in the end.

Another lesson that could be drawn from Coburn and Russell's study (2017) is principals might have their own agenda to implement district policies. As the findings show, school leaders in Region Z configured coaching resources from the district differently, for example, often assigning coaches to perform non-academic tasks. As a result, teachers in those schools had only fragmented interactions with coaches and the depth of interaction was much more limited. Furthermore, school leaders at both districts did not always convey messages congruent with the district aims, which discouraged teachers from implementing the curriculum in their classrooms. For instance, one principal focused on test preparation strategies that were not the aims of the innovative curriculum. In short, school leaders might interrupt or strengthen the efforts of the district to support the development of teachers' professional communities at the school site. Considering the fundamental roles of school leaders in changing instructional practices (Spillane, Hallett, & Diamond, 2003), further studies are needed to examine interactions between reform policies, school administrators, and teachers' change.

Professional development and curriculum change

One underlying message of the preceding studies is thoughtful professional development greatly facilitates reform. That is a time-honored lesson. Early curriculum implementation theorists already noted that teachers can only fundamentally change their practices when they receive extensive professional development to upgrade their attitudes to make them congruent with the reform ideas (Stenhouse, 1975). In practical terms, "high quality professional development is a central component in nearly every modern proposal for improving education" (Guskey, 2002, p.381). Professional development is considered by teachers as among the most promising and most readily available routes to growth on the job (Fullan, 1991; Guskey, 2002). Effective professional development can catalyze teachers' change in their classroom practices, in their attitudes and beliefs, and in students' learning outcomes (Guskey, 2002).

Cohen and Hill's (2000) study evidenced that academic content-oriented professional development could positively enhance teachers' reform knowledge and practice. They used self-report data gathered from 250 elementary schools, including 975 teachers of mathematics in California in 1994, to explore relationships between teachers' opportunities to learn about reform, their knowledge and practices, and students' achievement, 9 years after the state introduced mathematics education reform standards. To discern how teachers' professional learning might predict teachers' reform practices, the dependent variables – teachers' practices – were measured by 14 Likert-type items, and teachers' opportunities to learn – that is, "time in student curriculum workshops," "time in special topics workshops," and "past framework opportunity to learn" – were input as independent variables. Several sets of least-squares regressions were run. The study confirmed the hypothesis that the more opportunities teachers had to learn the new mathematics and reformed instructional practices, the greater their practice would move in the direction that the state policy had proposed.

An especially informative finding was that professional development designed to help teachers learn the mathematics curriculum that students used, that is, "grounded in academic content" (p.330), was more likely to produce constructive effects in improving teachers' practice and student performance. However, the reality was bleak: Most professional development programs for most teachers in California did not change teachers' practice or students' achievement (Cohen & Hill, 2000). Two conditions might help foster effective professional development for teachers: Increasing teachers' participation in making school policy and planning professional development activities, and constructing and maintaining a stable school community (Desimone, Smith, & Phillips, 2007).

If reformers expect to achieve desired curriculum reform outcomes, it is particularly significant to offer coherent professional development experiences to teachers (Penuel et al., 2007). Cohen and Hill (2000) recommended that coherent and effective professional development should integrate curriculum for students, provide teachers opportunities to learn the curriculum, focus teaching on learning, and align curriculum, assessment, and teaching. This is easier said than done: In their study, only 15 teachers, or 20%, had effective, coherent professional learning opportunities. Apparently, orchestrating professional development in a coherent fashion represents a daunting task for U.S. public education, which is "a non-system" (p.331) per se. Oftentimes teachers are engaged in fragmented, short-term, and procedural activities that lack in-depth study of the school curriculum or thoughtful plans to improve teaching and learning. One possible way to address this shortcoming is promoting on-the-job learning in conjunction with formal professional development. It appears that on-the-job learning can have a significant impact on teachers' instructional change (Parise & Spillane, 2010).

Disengagement of parents in curriculum reform

Believing that parents have a critical role in children's mathematics learning, researchers have called for increased parent participation and involvement in mathematics education reform and school decision-making (Peressini, 1998). In reality, parents are generally distanced from schools, and "reformers in mathematics education have not extended careful consideration to parents and their interests" (p.562). As a consequence of this distancing, parents fail to understand the need for mathematics education reform and can hardly make sense of the rationale behind the suggested reform programs, not to mention having any ownership in implementing the reform curriculum (Peressini, 1998). The failure of the majority of parents to make their voices heard in the reform might leave a minority of parents to pursue their own self-interests in schools.

In particular, parents of lower socioeconomic and educational background might be more disadvantaged in mathematics education reform. By interviewing ten African American parents from a low-income neighborhood, Remillard and Jackson (2006) discovered that those parents had little knowledge of the innovative approaches to mathematics teaching and learning and, hence, were disempowered to take part in the discourse of reform. The implementation of a reform-minded mathematics curriculum essentially prevented those parents from taking a more active role in supporting their children's learning.

Today, what Peressini (1998) claimed still stands sound: "[E]fforts to involve parents must be continued if educators are to successfully implement a vision of reform-based mathematics education" (p.566). Policymakers, educators, and researchers should strive to fulfill this agenda.

Note

1 Mason and colleagues (2005) refuted the study by Ogawa et al. (2003) as a "pejorative portrait" of the school district's standards-based reform efforts. In a 1998 mid-year anonymous evaluation of the standards-development process, they surveyed 374 teachers and school representatives from the district in Ogawa et al.'s study. The reform efforts were rated positively. Furthermore, a difference-of-differences econometric model was employed to estimate the effects of the implementation of standards and criterion-referenced tests at the district's elementary schools from 1999 to 2002 by using data from the California Department of Education. In summary, ten out of 16 positive and significant effects were found in Grades 2–5 mathematics achievement tests. Furthermore, in Grades 2 and 3 in 2002, students with 3 and 4 years of exposure to the program, respectively, experienced positive to significant (3 out of 4) reading and mathematics effects. On the one hand, this study suggests that the district's standards-based reform generated positive outcomes; on the other hand, it may well corroborate Ogawa et al.'s claim that the district's reform did indeed take two forms simultaneously: appearing both rational and ceremonial.

3 Research design

This study employed a qualitative case study approach. "A qualitative case study is an intensive, holistic description and analysis of a bounded phenomenon" (Merriam, 1998, p.xiii). By *qualitative*, it means that the inquiry is essentially concerned with making sense of things within their context. Put another way, qualitative research is:

> ...A situated activity that locates the observer in the world. It consists of a set of interpretive, material practices that make the world visible. These practices transform the world. They turn the world into a series of representations, including field notes, interviews, conversations, photographs, recordings, and memos to the self. At this level, qualitative research involves an interpretive, naturalistic approach to the world. This means that qualitative researchers study things in their natural settings, attempting to make sense of, or interpret, phenomena in terms of the meanings people bring to them.
>
> (Denzin & Lincoln, 2005, p.3)

Qualitative research purports to understand the process of a phenomenon by "investigating the viewpoint of those studied" (Becker, 1996, p.58), while attending to researchers' reflexivity in the construction of meanings.

A qualitative inquiry will be better suited to make use of the case study strategy, if it meets the following criteria: (1) It investigates a contemporary phenomenon within its real-life context; (2) when the boundaries between phenomenon and context are not clearly evident, and in which; (3) multiple sources of evidence are used (Yin, 2003). That is, the case study strategy has its particular advantages when the phenomena in question are contemporary and interwoven with the context, and when a holistic and in-depth investigation is called for.

In light of Yin's (2003) recommendations, we opted for the qualitative case study approach for this research. Three particular reasons informed this methodological decision. Firstly, the mathematics curriculum reform is still going on in China's schools and remains a hot topic in contemporary educational discourses. Administrators, teachers, parents and students are living through this transformation of curriculum and teaching day in and day out.

Secondly, an in-depth understanding of curriculum reform cannot be detached from its sociocultural contexts. Curriculum is not immune to the historical, social, cultural, and political particularities of the society in which it is designed and administered; on the contrary, it is precisely a social cultural product (Cornbleth, 1996). A study of curriculum reform has to pay due attention to both the phenomenon and the underlying social cultural context.

Thirdly, technically, case study has the flexibility and viability of grappling with the complex nature of curriculum reform by utilizing multiple sources of data. Case study is characterized by its triangulating research strategies. Triangulation, as Stake (1995) argues, is considered "a process of using multiple perceptions to clarify meaning, verifying the repeatability of an observation or interpretation...to clarify meaning by identifying different ways the phenomenon is being seen" (p.241). A case study often draws on multiple sources of data as deemed relevant to the study, such as documents, archival records, interviews, direct observation, participant observation, and physical artifacts. No single source of data has a complete advantage over others; rather, they are complementary to one another. Multiple sources of data warrant the trustworthiness of research findings.

There are three kinds of case studies: Exploratory, explanatory, and descriptive (Yin, 2003). Exploratory and descriptive case studies focus mainly on "what" questions and may be the initial steps into social research that needs deeper investigation. However, explanatory case studies are often used in answering "how" and "why" questions, from which causal relations of phenomena can be derived. This research study was more exploratory and descriptive in nature, which meant depicting the implementation process of the reform and people's responses to it in the case schools. Because of this designed self-limitation, the study yielded no causal explanations for the phenomena detailed in the book. Though, we do present inferences on several occasions.

Yin (2003) makes explicit five key components of a case study: The study's questions; its research hypotheses or propositions, if any; unit(s) of analysis; the logic linking the data to the propositions; and the criteria for interpreting the findings. It is worth noting that not every case study, for example, exploratory studies, has a research proposition, nor is it necessary. The unit of analysis can be an individual, an organization, a program, or a decision, which defines the boundary of a case. The last two components, the data analysis steps, are the least developed in the case study methodology, however. This study treats the school as the fundamental unit of analysis to illuminate the overarching research question. We did not start with any preconceived hypotheses or propositions, in line with the tradition of qualitative research. Rather, we let data, which were interpreted within the conceptual framework of structural symbolic interactionism, direct exploration (Stryker, 2008). As a modified version of George Mead's classic symbolic interactionism (Blumer,

1969), structural symbolic interactionism is premised upon the revised assumption that society shapes self, which shapes social interaction. Stryker (2008) posits that "although society emerges from social process, organized society exists before the appearance of all new members" (p.19). This view sees social structures at all levels as considerably durable, resistant to change, and capable of reproducing themselves. Social structures both facilitate and constrain human interactions in social networks. Large-scale social structures, such as class, age, gender, and ethnicity, realize their effects on interpersonal relationships through the operation of intermediate social structures, such as differentiated groups, communities, schools, and institutions. These intermediate social structures shape the content of self and its organization. Thus, while placing an equal emphasis on human meaning making, proceeding and mediating acting, the revised frame, compared to Mead's view (Blumer, 1969), underscores the impact of meso-level structures upon social interactions and role relationships of members.

The school organization constitutes one typical meso-level social structure. In light of structural symbolic interactionism, on the one hand, school institutions, such as policy mandates, norms, and regulations, are made possible via the interpretations of teachers, school leaders, and parents in social and cultural interactions. Indeed, people in schools do not act in automatic homogenous ways, as if they were programmed to execute the rigid mechanism of school bureaucracy. On the contrary, they derive individual meanings out of rules and norms, negotiate their identities, and play their roles accordingly. On the other hand, once the structural entities are established, they promote preferred interactions and behaviors, and constrain undesired ones in the organization. Thus, teachers' sensemaking and agency are bounded within the particular local structures.

It is important to state here, our basic belief is that theories cannot be treated as acultural truth. Instead, they are situated in the broader sociocultural and political context. Hence, we referenced to the above Eurocentric theoretical lenses, but we did not confine ourselves when these lenses failed to decipher phenomena deeply rooted in Chinese culture. Essentially, "a Chinese individual, far from being a distinct and separate *individuum*" (Sun, 1991, p.2), does not "belong[ing] to himself or one particular person, but is shared by all significant others" (p.40).

Understanding Chinese schools and educational phenomena can hardly eschew this deep, defining cultural structure that superimposes (or undergirds) other macro- and meso-social structures. Especially when it comes to the persistent examination obsession that Chinese educational reform attempts to dismantle, this deep structure is more capable of shedding light. Performing well in school is not a private, individualistic business of the child; instead, it concerns not only *parents' not losing face* (*bu diu fu mu de lian*; *bu*=not, *diu*=lose, *fu*=father, *mu*=mother, *de*=of, *lian*=face) when compared with others, but *surpassing other similar heads in the field* (*chu ren tou di*; *chu*=pass, *ren*=person, *tou*=head, *di*=earth) to make a decent living and ultimately honor the blood and ancestry (*guang zong yao zu*; *guang*=light, *zong*=ancestry, *yao*=glorify, *zu*=ancestry). Otherwise, it humiliates both oneself and one's family.

Context

Site selection

The study was carried out in two city elementary schools in one central district of a northeastern city in China[1]. We intentionally did not situate this study in economically developed and resource-ample major cities that, to our belief, hardly reflect China's ordinary educational conditions. In addition, the study was conducted in public city schools instead of rural or suburban ones. It should be noted that city or urban schools in China are different from their counterparts in the U.S. City schools in the U.S. often convey an image of minority children, lack of resources, and poor academic achievement. However, China's city schools have much more resources and are more affluent than rural or suburban schools. In this sense, China's city schools are most comparable to the U.S. suburban ones.

The City had a population of over 4 million. It was economically below average. In 2019, its gross domestic product (G.D.P.) per capita ranged between 5,500 and 6,000 Chinese yuan (between 800 and 850 U.S. dollars (U.S.D.); 1 U.S.D. roughly equaled 6.5 yuan), ranked between 350th and 450th among 611 Chinese cities[2]. An elementary teacher's salary was around 3,000 yuan per month. Unlike schools in major cities that tend to attract college graduates, elementary teachers in this city were mostly recruited from secondary-level normal schools.

The City was made up of three prefecture-level districts and nine counties. The study was carried out in the Red Pebble District, the central district of the City, which used to be the location of the city hall. The District consisted of about 50 public elementary schools and more than 25,000 students were enrolled there.

We based our fieldwork in two urban elementary schools, Merits School and Pioneer School. These two schools were chosen owing to their established academic reputation in this City. In accordance with the administration of Chinese local government, schools in China consist of four tiers, schools in China consist of four tiers: The ones within the jurisdiction of the Provincial Education Bureau, the ones within the City Education Bureau (C.E.B.), the ones managed by the County or District Education Bureau (D.E.B.), and the ones managed by the Township Government.

Figure 3.1 displays the organizational structure and locates the two schools (Merits School and Pioneer School) in this hierarchy.

Merits School was directly managed by the Red Pebble D.E.B. That is to say, the school was held accountable to its immediate supervisor, the D.E.B., and subject to its guidance and oversight. Pioneer School was within the

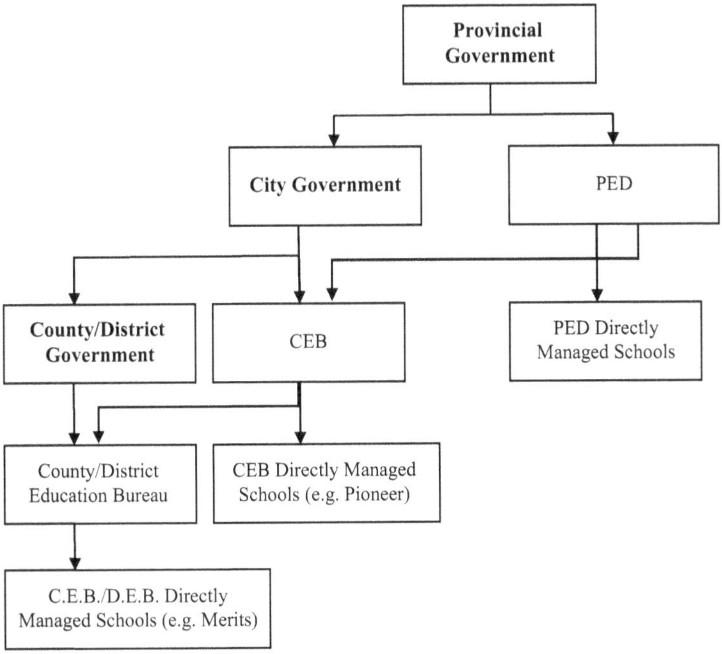

C.E.B.= City Education Bureau, D.E.B.= District Education Bureau, P.E.D.= Provincial Education Bureau.

Note: [a]Pioneer School was downgraded in 2006 and was directly managed by the Red Pebble DEB at the time of this study.

Figure 3.1 General managerial hierarchy of Chinese schools.

jurisdiction of the C.E.B. from early 1980 to 2006. The then principals held the same level of official rank as the head of the D.E.B. Compared to Merits School, Pioneer School enjoyed a higher degree of autonomy during those years, and teachers had more advantages and opportunities in advancing their career. Even though both were located in the same district, and in principle Pioneer School should have conformed to the management of the D.E.B., most often it was the case that the school disregarded administrative orders and possible educational campaigns initiated by the D.E.B. In 2006, Pioneer School was downgraded to the same level as Merits School, and as a result the school lost its previous privileges.

Merits School

People. Merits School was founded in 1966. In total, Merits School had 109 staff members: Ten males and 99 females. Six out of ten males and 97 out of 99 females were in full-time teaching (not administrative) positions. The school was departmentalized according to subject areas. Chinese and mathematics were the two main subjects (*zhu ke*; *zhu*=main, *ke*=subject) in the school. Other content areas, such as science, arts, and English, had marginal status and were habitually referred to as "para-subjects". There were 36 mathematics teachers (three of whom were male), 36 Chinese teachers, and eight English teachers. Every mathematics or Chinese teacher taught one particular class, while an English teacher was responsible for two different classes. Teachers at Merits School specializing in one of the main subjects were also required to teach one or two para-subjects like science, social studies, or moral education, because their weekly teaching load was considered less than 30 hours. Those who assumed the position of classroom director (*ban zhu ren*; *ban*=class, *zhu ren*=director) did not have to take on additional para-subjects. Other non-teaching personnel worked in positions such as I.T. support, archiving, or administration.

There were 36 classes at Merits School. The size of the classes ranged from 42 to 70, amounting to 2,140 students in total, with on average 60 students per class. See Table 3.1 for the number and size of classes at Merits School.

Every student was equipped with a desktop computer, and teachers were supposed to deliver teaching via the computer and the television projector; the attempt was aborted shortly thereafter.

School organization. The highest administration of Merits School consisted of the Principal and two vice principals (V.P.s), one in academic and the other in moral education. The Principal was responsible for the overall operation of the school. The academic V.P. took charge of teaching and learning matters. The V.P. in moral education focused on issues related to students' morality and behaviors. On the periphery of the organization, Merits School also had one party secretary, in charge of affairs of the school Chinese Communist Party. Merits School's organizational chart is depicted in Figure 3.2.

- Under the leadership of the academic V.P. was the Teaching Guidance (T.G.) Office that constituted the most important secondary hierarchy of the school administration. Merits School had one teaching director (T.D.), and ten associate teaching directors (A.T.D.s):
- A.T.D. 1 oversaw library and archives.

Table 3.1 Class sizes at Merits School

Class	*Grade*					
	1	*2*	*3*	*4*	*5*	*6*
1	65	63	63	64	63	64
2	63	62	61	63	61	60
3	60	62	62	60	59	58
4	61	61	58	56	61	58
5	47	54	52	51	52	51
6	70	69	69	69	46	42

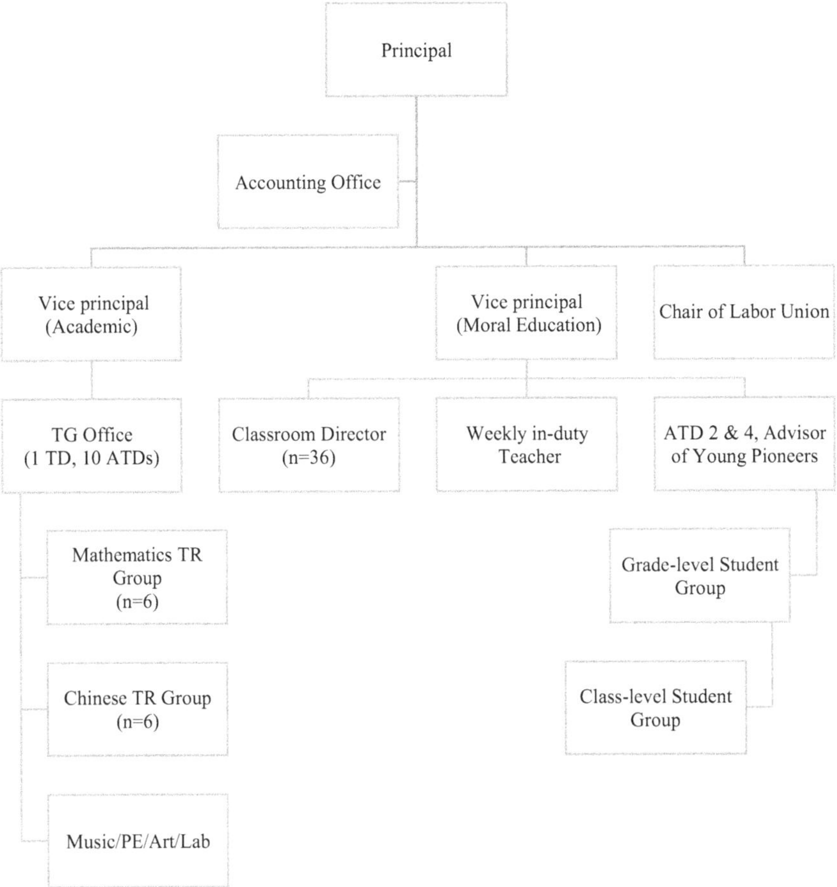

Note: A.T.D.= associate teaching director, T.D.= teaching director, T.G.= teaching guidance.

Figure 3.2 Merits School organizational chart.

- A.T.D. 2 was responsible for moral education, school hygiene, and safety.
- A.T.D. 3 led teaching and research in Grades 5 and 6 for Chinese and English.
- A.T.D. 4 managed the organization of Young Pioneers, and directed Grade 2 mathematics.
- A.T.D. 5 was charged with music, physical education, art, health, technology-assisted education, and experiments.
- A.T.D. 6 orchestrated overall teaching and research, and Grade 3 mathematics.
- A.T.D. 7 was in charge of comprehensive practices of students and Grade 6 mathematics.

- A.T.D. 8 focused on Grades 3 and 4 Chinese.
- A.T.D. 9 monitored teaching and research of Grades 1 and 2 Chinese.

A.T.D. 10 took care of Grades 1 and 4 mathematics and compiled teaching evaluation materials.

Under the T.G. Office were grade-level teaching research (T.R.) groups, hosted in separate offices by subject areas. Each T.R. group selected one outstanding teacher as the team leader, who was not counted as a school administrator. Merits School had six mathematics T.R. groups or offices, six Chinese, one science, one English, one art, one music, and one physical education. It was evident that mathematics and Chinese were the major subjects, and all other areas had less important status at Merits School.

The V.P. in moral education at Merits School directly oversaw 36 classroom directors. Each classroom was assigned one teacher as the classroom director. A classroom director was a regular teacher with extra responsibilities. He or she took full responsibility for the class and was often a mathematics or Chinese teacher, rarely a teacher in a para-subject area. The other main subject teacher would assist or collaborate with the classroom director. Being a classroom director was often the recognition of one's dedication, teaching excellence, and sense of responsibility.

Position responsibilities. Since schools in China are subject to the control of the nation and responsibilities of major positions are similar across the country, here we use Merits School as an example to illustrate the responsibilities of major positions. As detailed in the Merits School's Policy Collection, which contains 114 regulations over 124 pages, the Principal has eight broad duties, namely

(1) Being in charge of the overall operation, organizing and leading staff members to conscientiously carry out national guidelines, policies, and plans.
(2) Making school development plans, and establishing and enforcing school regulations and rules.
(3) Organizing and leading school administrators in the beginning of the semester to make the year plan in line with the D.E.B.'s yearly goals to which the school will be held accountable and the school reality.
(4) Being responsible for building up the synergy of the school leadership and continuously leveling up capacity of the whole school.
(5) Being responsible for developing the team of teachers via a variety of activities focused on ethics and specialty.
(6) Presiding over school-wide and administrators meetings.
(7) Being in charge of school human resources and year-end teacher appraisals, hiring new teachers, and awarding or punishing teachers.
(8) Examining and approving all school expenditures.

Formally residing at the interface of the school and the state, the Principal should perform the role of the state delegate by echoing and executing official mandates and policies. Personnel, budgets, and other material resources – the lifelines of an organization – are controlled by the Principal. The Principal is the sole decision-maker in the school in terms of whether or not to hire a person, whether or not to accept a student from another school region, how to spend funds, or whether or not to rent a school space to external business people, and the like.

The responsibilities of the academic V.P. have nine aspects:

(1) Assisting the Principal to lead teaching and research at the school.
(2) Guiding and orchestrating the T.G. Office to put forth school plans, and encouraging and examining the implementation of the plans.
(3) Detailing teachers' responsibilities and specifying the criteria used to evaluate the performance of teachers.
(4) Designing the school curriculum schedule in line with national curriculum outlines or standards.
(5) Aiding the Principal in hiring new teachers and appraising teachers.
(6) Making school teaching research plans, selecting research topics, advising teaching research, and organizing teaching research training.
(7) Being in charge of improving teachers' professional capacity, organizing teachers to participate in training and continued education.
(8) Being responsible for purchasing instructional instruments, books and materials.
(9) Being in charge of teacher attendance.

The academic V.P.'s central work, as shown above, is focused on the arrangement and oversight of the school's teaching, learning, and teaching research activities, the technical core of Chinese elementary education.

Under the academic V.P.'s supervision, the T.G. Office takes on the executive responsibilities. The duties of the T.D. are delineated as follows:

(1) Planning teaching tasks, making daily schedules, and arranging the curriculum and extracurricular schedules.
(2) Maintaining teaching norms[3] (*jiao xue chang gui*; *jiao xue*=teaching, *chang gui*=rules of routine; generally consisting of lesson planning, instructing, student work, tutoring, and assessment) and orderliness, and assisting the principal in assigning teachers to teaching posts and appointing classroom directors.
(3) Leading T.R. groups, directing T.R. and collective lesson preparation activities, summarizing and disseminating outcomes, and organizing teachers to learn and develop professionally.
(4) Dealing with everyday affairs, for example, student enrollment, class allocation, student demotion or promotion, suspension, transfer, dropout, and graduation, coordinating class exchanges and substitution, ordering and disseminating textbooks, and planning purchases of instructional instruments, apparatus, experimental equipment.
(5) Making detailed teaching evaluation criteria and conducting educational and teaching evaluations.
(6) Creating student registers and teacher's professional records.

(7) Frequently sitting in T.R. groups and classrooms, observing teachers' teaching performance, cultivating backbone teachers (*gu gan jiao shi*; *gu*=bone, *gan*=stem, *jiao shi*=teacher)[4], mentoring young teachers to improve instructional skills, collecting suggestions and requests from teachers and students, and helping the principals to solve issues in teaching.

The T.D. and A.T.D.s are responsible for overseeing day-to-day teaching affairs and ensuring the quality of teaching and learning of the whole school. They carry out regular evaluations of teachers.

Classroom organization. Classroom organization in China mimics the organization of the school system, which is modelled on governmental institutions. The classroom governing structure is made up of two levels: The classroom teachers and the student governing body. Each classroom has one teacher as the classroom director. The role is often taken by a mathematics teacher or Chinese teacher, rarely by a teacher in a para-subject area; the other main subject teacher will assist or collaborate with the classroom director. As the Policy Collection specifies, the classroom director should:

(1) Make long-term and semester plans and conduct moral and character education in the class, stress the cultivation of benign moral attributes, and good learning, working, hygiene, and behavioral habits.
(2) Construct the classroom organization, that is, elect student committee members, group leaders, and subject representatives within two weeks after the school year opens.
(3) Hold at least one meeting with student officials and have them become the leaders of the class.
(4) Plan a rich variety of forms and contents of meetings.
(5) Advise on the use and protection of student desks, chairs and other classroom properties, and enforce detailed rules on students' behavior during lesson breaks and school dismissal.
(6) Carry out education in safety and the law regularly to prevent accidents from happening;
(7) Care for every student, manage students in a strict manner, and be able to use different approaches to educating and transforming laggards.
(8) Keep in touch and collaborate with subject matter teachers frequently, have an all-around knowledge of every student, and make regular evaluation of students.
(9) Assign students to clean the classroom and allocated campus areas.
(10) Care about the physical health of students, protect their vision, and rotate their seats once a month.
(11) Contact parents regularly, win over their support and cooperation, and hold one teacher–parent conference per semester.

(12) Accomplish any task assigned by the school in a complete and timely manner and to a high standard.
(13) Fill out the director's working journal accurately.
(14) Behave responsibly, practice what they preach, and be an exemplary role model for students in every respect.

In simple terms, she or he is held accountable for students' behavior, learning habits, safety, character, discipline, academic performance and the like. For instance, the classroom director should escort his or her students to leave school at noon and evening school dismissal on a daily basis. The classroom director needs to keep in touch frequently with parents, give timely feedback to them, and, sometimes even reprimand or educate them. Classroom directors should arrive at the school 10–20 minutes before other teachers in the morning and at noon, and leave later in the evening. They need to watch students' behavior throughout the day and be present at every convened event, like the morning intermission exercise, raising the national flag, and standing in line at school dismissal.

On an 80-point scale, classroom directors at Merits School are appraised on three broad categories, seven subcategories, and 32 specific items. To name a few, classroom directors should attend on time the school training on classroom management issues, otherwise 2 points will be deducted per absence, 0.5 point will be deducted per instance of sick leave, 0.5 point per instance of tardiness; if there is litter in the hallway outside the classroom, the corresponding classroom director will lose 1 point.

There is a small monthly stipend for the classroom director in the amount of 50 Chinese yuan in addition to the salary. Some parents may invite the teacher out for a thank-you dinner. It is also more than occasional, as an underground norm, that some parents may give gift cards to the classroom director for holidays, in the hope of having teachers take better care of their children, for example, calling them more often to answer questions, scolding them less in class, or assigning a good seat in the front rows.

Regarding the student part, there is a rather sophisticated student governing system. The classroom is managed by classroom officials (*ban gan bu*; *ban*=class, *gan bu*=official). One student, usually one who performs well above others behaviorally and academically, is nominated (sometimes selected by the whole class) as the class chairperson (*ban zhang*; *ban*=class, *zhang*=head). He or she is a kind of *teacher assistant*. Under the chair is one or two vice chairs, and five or more committees responsible for academics, hygiene, classroom discipline, physical exercise, and recreational or performance activities, respectively. Each committee has one head. Under the academic head are subject representatives, often appointed by the subject area teacher, mainly in English, Chinese, and mathematics. In the morning, the representative collects the homework books for the corresponding subject and sends them to the teacher's office. They also write down and submit the names of these who do not turn in homework. Before the beginning of the class, the representative goes to take back students' homework books from the day before.

The head of classroom hygiene is to ensure that his or her peers complete their duty to clean the classroom. Every day, four or more students are charged with cleaning and refreshing the classroom floor in the morning, disposing of trash, arranging tables in line, and erasing the blackboard after each lesson. In addition to the classroom duty, each class may be allocated a small portion of the campus to take care of. Groups of students need to rotate to perform that task. In a similar vein, the head of classroom discipline helps maintain classroom order, and the heads of physical education and recreation help organize relevant activities. These two positions are more or less nominal. Students' seats are organized into four columns and seven to eight rows. Each row is made up of four desks, each with two students. Generally, each column is one group and this group tends to select one student as the group learning leader who is to regulate and promote learning.

This arrangement promotes the self-management, participation, and independence of students to some extent, but it also puts children in an overt hierarchy . This mechanism of power operation stratifies the managing from being managed, the controlling from being controlled, and the good, docile students from the "bad" black sheep. More often than not, rare resources like the teacher's attention, awards or honors from the school are allocated to student officials rather than ordinary students.

Pioneer School

People. Pioneer School was founded in the 1940s. It had 112 staff members (three were affiliated to the school, but did not show up), consisting of 23 mathematics teachers, 35 Chinese teachers, and 7 English teachers. Other staff included teachers of music, arts, physical education, social studies, science, and an archive keeper, and full-time administrators. In total, there were 14 males, 8 in full-time teaching posts, and 95 females. Historically, more parents of children who attended this school worked for the government, were wealthy, and had strong *guan xi* (connections) than those of other schools in this City.

There were 35 classrooms in Pioneer School. The size of classes ranged from 56 to 73, amounting to 2,161 students in total, and with an average of 62 students per class. Pioneer School was also affected by the "Year of the Tiger effect" (only five classes in Grade 5). The Principal mentioned the superstition of the local people who believed that it was not lucky to give birth to a baby in the Year of the Tiger. The smallest class size still exceeded the C.E.B.'s top limit of 48 by 8 students. See Table 3.2 for the number and sizes of the classes.

School organization. Pioneer School had an organizational structure similar to Merits School, as shown in Figure 3.3, but several elements of the organizational structure were different. First, Pioneer School had been allowed to set up an office to coordinate between the Principal, teachers, and the district. The office remained after the school was downgraded to the district level. The second difference was that Pioneer School only appointed three A.T.D.s,

Table 3.2 Class sizes at Pioneer School

Class	*Grade*					
	1	*2*	*3*	*4*	*5*	*6*
1	64	65	59	64	64	63
2	58	56	56	58	58	59
3	60	57	58	59	59	59
4	56	56	57	60	61	61
5	63	62	67	62	65	–
6	68	70	68	73	69	67

which was more common at most schools in the district. One A.T.D. was in charge of mathematics, one was in charge of Chinese, and the third was in charge of the rest of the subject areas. Thirdly, there were only three mathematics T.R. offices (or groups), which were the Grades 1–3 T.R. group, the Grades 4–5 T.R. group, and the Grade 6 T.R. group. Because most mathematics teachers, except Grade 6, taught two classes, Pioneer School had fewer mathematics teachers than Merits School.

Classroom organization. Pioneer School's classroom organization was similar to that of Merits School. At the Pioneer School, students were evaluated taking each individual classroom as the unit of inspection. Students' orderliness, behavior, and hygiene habits were taken into account. Table 3.3 was the class-by-class result of the school inspection in the 15th week of the school year.

The school used weekly inspection to help students form desired classroom habits. Points were taken away if students were found in violation of school rules. In Table 3.3, note that 1.5 points under the subcategory, within building discipline, were deducted from Grade 6 Classes 1 and 3; Classes 4 and 5 lost 1 point each.

Because the size of classes was too large, seat allocation was tricky. For a while, it had been a highly contentious issue at Pioneer School. Parents reported to the C.E.B. the unfairness in seat allocation. The Principal finally designed a rotating plan and seats were allocated according to students' heights. Every two weeks, four rows would rotate horizontally, so would columns vertically. The head of the C.E.B. and the Principal occasionally visited the classrooms unannounced and asked students to stand up to make sure there was no hidden deal. Perhaps, this side story might show how schools were complicated organizations in China.

Accessing the field

In China, there were no laws or rules specifying what a researcher should go through or abide by when conducting a study in a school or involving children.

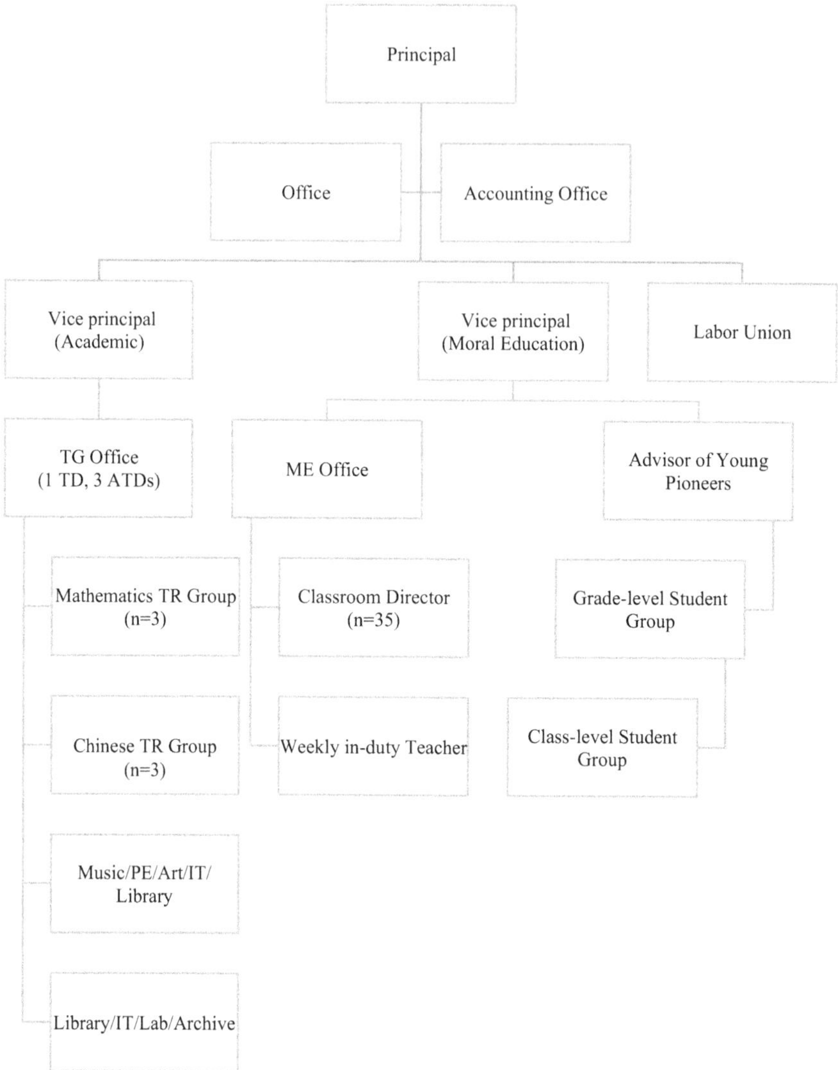

Note: M.E.= moral education, T.G.= teaching guidance, T.R.= teaching research

Figure 3.3 Pioneer School organizational chart.

It was up to the individual school's discretion. In circumstances when formal rules are not in place, connection or *guan xi* in Chinese society is useful. *Guan xi* is built upon amicable, reciprocal, interpersonal terms, reflecting the closeness and ubiquity of mutual reciprocity (*ren qing*; *ren*=human, *qing*=feeling), and is key to the success of many endeavors at a local level in

Table 3.3 Weekly inspection of cultivation of classroom habits

Class	*Orderliness of personal items*	*Eye exercises*	*Taking the bus*	*Queuing at school dismissal*	*Discipline*			*Hygiene*			*Extra*	*Total*
					within building	*within room*	*intermission exercise*	*room*	*school zone*	*personal*		
6.1					1.5			1				97.5
6.2												100
6.3					1.5				1			97.5
6.4					1							99
6.5					1							99
6.6									1			99

Note: 6.1 means Grade 6 Class 1.

China. With *guan xi*, iron rules can be bent and bypassed; without *guan xi*, it will be strenuous, sometimes futile, to confront with tedious bureaucratic procedures. We reconnected with several previous colleagues and friends. Ms. Sung,[5] a previous acquaintance of mine and middle school teacher, was particularly supportive after learning our research needs. Together we identified four schools as possible choices based on the criteria of location, academic reputation, and student population. That is, the school should be in an urban area, as discussed previously; it must have a fine reputation in academics and reform implementation according to the judgment of officials in the local bureau and parents at large; and it should enroll multiethnic students instead of being a school that has little or no ethnic diversity. Some schools favor only one particular ethnic group in line with Chinese ethnic polices, for instance, schools for ethnic Mongolian, Tibetan, or Uighur children. Two schools that best fit our profile were finally singled out.

We summarized our research agenda via email respectively to Principal Yong at Merits School and to Principal Huang at Pioneer School. Both of them were promoted to their current positions six months earlier. Following the emails, we talked to them over the phone. Principal Yong sounded lukewarm about our proposal, but did not refuse either. Ms. Sung told us later on that Principal Yong was actually a very easygoing person if a good relationship with him could be nurtured. Principal Huang was very enthusiastic and candid, though he was somewhat suspicious of the veracity of our study in the beginning. He asked frankly whether we would conduct this study for real, or simply need his cooperation to falsify an experience in his school. "If you need something fake, let's do the fake way. If you need something real, let's do the real way," Principal Huang said. It is understandable, since it would not surprise him if we were there only for proof of an experience.

At Merits School, Principal Yong introduced T.D. Wang of the T.G. Office during the first meeting. T.D. Wang had headed the T.G. Office for 11 years, was also a Grade 6 mathematics teacher, and seemed to hold a prominent place in the school. Both of them singled out three mathematics teachers, Teacher Zhang from Grade 1, Teacher Feng from Grade 2, and Teacher Hong from Grade 5. They did not recommend that we "disturb" any teachers from Grade 3 or Grade 6 in the study, since teachers in both grades were painstakingly preparing for the upcoming district-wide Uniform Examination (*tong kao*; *tong*=uniform, *kao*=test) on July 15; it was their first and only priority. Other grade levels also needed to take the test, but those students were monitored by their own teachers during the test, and their answer papers graded only at in their own schools. Grade 3 and Grade 6 students, however, took the test under the supervision of teachers from other schools, and needed to submit their papers in sealed envelopes to the District T.R. Center.

At Pioneer School, Principal Huang appeared delighted to have us there. He even suggested that we hold a seminar with English teachers in the school and discuss how to learn English better. As at Merits School, he suggested that we not count on Grade 3 and Grade 6 mathematics teachers. Teacher Me, a nation-level backbone teacher, was recommended to me. She taught Grade 2 mathematics.

One week later, we invited the principals, Ms. Sung, and another two researchers to dine at a medium-level restaurant. Both principals joined me. It should not be interpreted as a form of bribery. Partly, it was to show a thankful gesture to them, since accommodating a researcher was not within their school duties. It would be fine if we pretended to be naive and did not treat them, but we would most likely be perceived as pedantic, or, not socially smart (*bu hui lai shi*; *bu*=not, *hui*=able, *lai shi*=deal with things). Partly, it was to forge a positive relationship or *guan xi* with the principals, the gatekeepers of the schools, who were critical to the study, since they controlled the resources and personnel that mattered to the scope and depth of information we could acquire. The dinner went nicely. "Just ask if you need us to provide any information," Principal Huang assured me.

Selecting participants

School administrators, mathematics teachers, and parents in the two schools constituted the key informants of this study. As discussed earlier, Chinese schools consist of four levels of hierarchy: The highest is the Principal, the second includes V.P.s, followed by the T.D. and A.T.D.s, and at the lowest level are ordinary classroom teachers. The Principal tends to make key decisions concerning the overall school operation. The academic V.P. and the T.D. are specifically responsible for academic affairs. Apparently, mathematics teachers, as the key arbiters of instructional content and practice (Cohen & Hill, 2000), are both the target of curriculum reform and the gauge of its outcomes. Parents, as important stakeholders in the educational enterprise, might also exert influences during the reform, so they were counted in as well.

We were conscious that how rich and true our data could be depended on the quality of our relationships with the school people. There was no better way possible but to spend longer and interact more frequently with our informants.

From the brief contact with T.D. Wang in the Principal's office of Merits School, she looked a little aloof and we were concerned how to get to know her more closely, since she was an important player in the school. It was cherry season. There were several cherry trees on campus. On one occasion, Teacher Zhang asked us to help her pick cherries because we were relatively taller. By chance, T.D. Wang and several other middle-rank administrators were also there. We readily helped everyone out. The cherry picking encounter enabled us to ease her guardedness and break the ice. Trust between us and our informants did not come naturally as time went by, but was cultivated by undertaking common concrete activities.

Teacher Zhang and Teacher Me served as our innermost core informants at their schools. We decided to spend the first two weeks in their offices. Our circle of contacts radiated around them. At Merits School, via Teacher Zhang, we developed good chemistry with all six Grade 1 mathematics teachers and subsequently had Teacher Rui join our study as one focal subject. In this way, we recruited 7 focal informants and involved three T.R. groups (n=18) in the study. At Pioneer School, Teacher Me connected us to the rest of the focal participants. She was a highly respected and recognized expert teacher in the school. Six teachers served as key informants in the study.

In order to illustrate the experiences of parents throughout the reform, four parents from each school were asked and agreed to participate. At Merits School, one parent was one of the six mathematics teachers in Grade 1, Teacher Tang, and her child was a Grade 6 student. We approached another parent on the street after school. He was a middle-level manager in a local company and his daughter was in Grade 2. The other two parents whose sons had just entered a local middle school were recruited in September. We had asked one administrator in the middle school to randomly pick them out. Both parents were unemployed and had no college education.

As for the four parent participants at Pioneer School, two were recommended by Teacher Me; they were female and in their early 30s. They were both public servants and affluent. The daughter of one of them was in Teacher Xiang's class, and the other's son was in Teacher Hua's class. The other two parents were also chosen at random via the middle school administrator. One was a housewife, laid off a decade before, and the other a beauty salon owner. Both had sons.

Table 3.4 displays the key focal participants from each of the school sites.

Data: sources, collection methods, management and analysis

The start list of data collection

We probed the informants with a "start list" (Huberman & Miles, 1983) of constructs that functioned as initial conceptual handles to guide our exploration. Perhaps, it is safe to claim that no inquiries start from the tabula rasa. Both our prior knowledge of Chinese society and our consultation of academic literature informed us of potential areas that we could focus on. The preliminary review into the literature suggests that studies concerning curriculum change need to accord due respect to the voices of administrators, teachers, as well as parents. Omitting any one of the three groups would render the picture incomplete. The literature brought to our attention important aspects such as administrators' attitudes toward the N.M.C.; formal and informal learning opportunities for teachers; teachers' change in knowledge, beliefs, and/or behaviors; student work; and parents' awareness of the reform; and the like. At the same time, from our vantage point as people who are Chinese, we were convinced that it was important to take a serious look into the Chinese school structure and institutions in place. The latter were often absent in the purview of scholars. These considerations led to a tentative checklist of areas of interest and possible data gathering methods to start with, as shown in Table 3.5.

Table 3.4 Demographics of focal participants

Merits School			*Pioneer School*		
*Participants****	*Education (initial degree)*	*Years of experience*	*Participants*	*Education (initial degree)*	*Years of experience*
Principal Yong	Normal school	(30)**	Principal Huang	Normal school	(24)
V.P. Yu	Normal school	22(32)	V.P. Yang	Normal school	(28)
T.D. Wang (G 6)*	Normal school	19	T.D. Zhi	Normal school	34
Teacher Rui (G 1)	Normal school	9	Teacher Me (G 2)	Normal school	33
Teacher Zhang (G 1)	Normal school	8	Teacher Tao (G 2)	Normal school	9
Teacher Feng (G 2)	Normal school	12	Teacher Jing (G 2)	Normal school	3(25)
Teacher Chen (G 3)	Normal school	12	Teacher Hua (G 3)	2-year college	3(15)
Teacher Hong (G 5)	Normal school	8	Teacher Xiang (G 5)	Normal school	14
Teacher Su (G 5)	Normal school	11	Teacher Quan (G 6)	Normal school	12
Four parents			Four parents		

*G 6=Grade 6.
**The number in parentheses refers to the total years of experience, including administrative experience.
***Other participants mentioned in the study include: Merits School – G 1: Teacher Wu, Teacher Tang, Teacher Ding, Teacher Zhu; G 2: Teacher Men; G 3: Teacher Xue, Teacher Nie; G 5: Teacher Fu, Teacher Nan; G 6: Teacher Zhou; A.T.D. Mei; Pioneer School – G 1: Teacher Jun; G 2: Teacher Wen; G 3: Teacher Yan, Teacher Ming; G 5: A.T.D. Teacher Chun.

Table 3.5 The start list of exploration

Conceptual handle	*Potential areas of interest*	*Content*		
		Document review	*Interview*	*Observation*
School information	History (e.g. when established)	School introduction	Administrator, teacher	N/A
	Values, purpose of education, emphases	Same as above	Same as above	
	Student information (e.g. expenditure per student or no. subsidies)	School record	Same as above	
	Teacher information (e.g. no. of teachers, experience, salary)	Same as above	Same as above	
	Administrative hierarchy	School introduction	Same as above	
	Schedule (e.g. curriculum)	School record		
	Emerging issues			
Reform overview: history, progress, issues	Initiation of the reform	Policy document	Administrator, teacher	N/A
	Purpose of the reform	Same as above	Same as above	
	Decision-making	School document	Administrator	
	Steps/strategic plans of reform	Same as above	Administrator, teacher	
	Periodical outcomes (when, how, and what)	Same as above	Same as above	
	Emerging issues			
Teaching	Class size (e.g. no. of students, gender)	Class roster	Teacher	Teacher
	Planning (e.g. individual, collective)	Class roster	Same as above	Same as above
	Instruction (e.g. differentiating)	Lesson plan	Same as above	Same as above
	Student work (e.g. importance, amount, format)	Student work	Teacher, parent	N/A
	Assessment (e.g. importance, frequency)	Student work Class ranking	Same as above	Teacher
	Emerging issues			

Teacher beliefs, knowledge, and practices	Experiences (e.g. preparation, teaching)	N/A	Teacher	Teacher
	Professional learning (e.g. collective planning)			
	Emerging issues			
	Beliefs (teaching, learning, learner)	Teacher journal Lesson plan, journal	Same as above	N/A
	Practice (e.g. classroom instruction)	Textbook, guides	Same as above	Teacher
	Knowledge (e.g. content, pedagogical)	N/A	Administrator, teacher	Same as above
	Daily work (e.g. time spent correcting work)	School plans, reports	Teacher, administrator	Same as above
Views of parents	Background (e.g. job, academic standing)		Parent	
	Knowledge, views of the reform and curriculum		Same as above	
	Out-of-school tutoring (e.g. Olympic Math)		Same as above	
	Emerging issues			

The table should not be seen as a pre-categorization of phenomena under investigation. It was simply a convenient roadmap and a reminder of important issues not to be overlooked. Nor did we report on all issues mentioned in the above table in the book. As a qualitative researchers, we remained sensitive to the exigencies in front of us in the field. We managed to avoid the tendency of taking things for granted because of the possible blindness caused by our familiarity with Chinese educational contexts. This point will be addressed in greater detail in the section on subjectivity reflection which follows in this chapter.

Data sources and methods of collection

Broadly speaking, we used multiple approaches to data collection. In qualitative research, relying on a singular data gathering method poses a danger of shrinking the richness and breadth of information. We tapped into both primary and secondary sources of data. The primary sources of data were obtained by means of interview and direct observation, coupled with memos and field notes. Interviews and observations are commonplace in a qualitative case study (Stake, 1995; Yin, 2003). The secondary sources were comprised of school archival records, local and national policy documents, participant artifacts (lesson plans, teaching materials, journals, and the like), and student work. Multiple sources of data allow for the convergence of findings and can effectively improve the authenticity of the study. For instance, to understand how teachers practiced, listening to the tales teachers told was only a partial representation; observing them instruct in class helped paint the picture better. Analyzing lesson plans, instructional materials, and student work and alike rendered the understanding more complete. We will describe these methods in greater detail in the following passages.

Interviews. One way of gathering data we employed was interviews. Semi-structured interviews were applied to all participants. We used a digital recorder to record all interviews.

The primary purpose of this technique was to enable us to delve into the internal world of the participants and gain a deeper understanding of their emic experiences and perspectives of the curriculum reform. As Merriam (1998) posits, highly structured interviews cannot afford a true participant perspective, but simply "get reactions to the investigator's preconceived notions of the world" (p.74). Semi-structured interviews allowed the participants to expand conversations, while ensuring that a few common questions were posed to all participants.

Sample questions for teachers were: What do you think mathematics is and why do we need to learn it? How do you understand the N.M.C. reform? How has it impacted on your teaching? Describe an ordinary day of your teaching before, during, and after the reform? Questions for principals and administrators included: How do you describe your school in terms of its mission and educational purposes? How did your school implement the

N.M.C.? What challenges have you encountered in carrying out the reform? How have you addressed these issues? As for parents, we asked questions such as: How do you describe the present school in which your child is enrolled? Could you tell me what you know about the new mathematics curriculum? Compared to the way you learned math, what do you find has changed with regard to the way your child learns math? How do you view such changes? How important do you think learning mathematics is to you and your child? What are the goals you have for your child in education? How do you ensure your child's success?

Equally important, interviews should be given multiple times, since one-shot interviewing can seldom capture the richness of a participant's perspective and life. Multiple interviews could "chart a person's path through a process" (Charmaz & Belgrave, 2001, p.318). Researchers could capitalize on emerging questions and themes out of the previous interviews to direct subsequent data collection.

We intended to interview each participant at least twice, and each interview should preferably last 90 minutes or longer. For teachers, it proved rare to find a considerable chunk of time, 90 minutes for instance, to take part in our interviews. Oftentimes, our conversations were cut off in the middle by classes, phone calls from parents, or impromptu school meetings. For us, encroaching on their after school time or weekends seemed inconsiderate. Moreover, several senior teachers suggested to us that we had better not do that. Thus, most interviews were conducted during school days, and lasted from 5 minutes of quick exchange of thoughts to 120 minutes of longer conversations. At Merits School, about 19 hours of interviews were conducted, which produced 196 pages of transcription. At Pioneer School, about 16 hours of interviews were conducted, which produced 147 pages of transcription.

Similarly, most parents could not spare the time or were less willing for a second lengthy interview. Each parent was only given one 60–90 minute interview. Concise follow-up phone calls to those parents were made when we needed to clarify some ambiguous points. The total time of parent interviews was about 7 hours and there were 42 pages of transcription.

Numerous phenomena that could not be preconceived were revealed via the semi-structured interviews. Take the staffing patterns of teachers as an example. There were at least three rotating schemes in place at both schools, namely, *big cycle* (*da xun huan*; *da*=big, *xun huan*=rotation), *intermediate cycle* (*zhong xun huan*; *zhong*=medium), and *short cycle* (*xiao xun huan*; *xiao*=small). That is, the teachers followed the students through the grade levels for several years. Respectively, if a teacher taught consecutively from Grade 1 to Grade 6, and then started over from Grade 1, that is called *big cycle*; if one only taught from Grade 1 to Grade 3 and then to Grade 1, or Grade 4 to Grade 6 and then to Grade 4, that was an *intermediate cycle*; in a similar vein, if the teacher rotated from Grade 1 to Grade 2 to Grade 1, Grade 3 to Grade 4 to Grade 3, and Grade 5 to Grade 6 to Grade 5, these were *short cycles*. The majority of teachers at both schools were arranged to teach big cycles, however, at the time of the study two male teachers (one was Teacher Su) taught only in the Grades 5-6-5 cycle, in part because they were able to better connect with and manage older students. Those teaching schemes were not random or trivial structures. To a great extent, multicycle rotating enabled teachers to acquire knowledge of the whole mathematics curriculum.

Observation. One possible limitation of interviews is that informants may gloss over their experiences or what one says one has done may not match what one really has done (Johnson & Turner, 2003). Observing the informants in situ is thus critical. Observation can capture the informants' actions in context. Observation can also expose the researcher to phenomena that otherwise might be foreign to him or her. As Whyte (1993) claimed, "I would not even have had the sense to ask [the questions] if I had been getting my information solely on an interviewing basis" (p.303).

Yielding quality data entails spending more time with the subjects, and being a visible member in their circles. Researchers coming and going in a fleeting fashion may undermine their credibility. As Teacher Wen, a Grade 3 mathematics teacher at Pioneer School, mocked:

> We had people come to do research. They stayed here one or two days, visited our classroom a few times, taped the lessons, and then were gone. That way, they could barely get anything meaningful. Real teaching is not that conspicuous. Plus, what they saw was very partial. We teachers could always put on a show for you. The 40-minute instruction is only one small piece of our everyday teaching job. It is these beyond the 40 minutes that matter to the quality of teaching...You have to stay with us, work with us day in and day out, and learn what we normally do offstage.
>
> (Informal Conversation in Grade 3 Group)

Her point of view deserves much attention from researchers. According to her emic perspective, even some cross-cultural studies conducted by reputable researchers, who endeavored to construe the "myth" of Chinese mathematics education, are in the danger of facing such charges. Videotaping and analyzing a few lessons might narrow the broader concept of *teaching* down to *instructing*. It runs the risk of dissecting the unity of teaching and isolating classroom instruction out of its inextricable context.

In this regard, consistent observation over a relatively longer period of time had particular merits. It allowed us to see the whole teaching activities with which mathematics teachers were engaged. Further, teachers showed their "undressed-up" classes to me. Teacher Me concurred, "Performing [for a teacher being observed] one or two times is easy. But it is impossible to play a show every day. Day by day you will see the real thing" (Informal Conversation in Grade 3 Group).

Let's give one example here to clarify the point. Preparing lessons ahead of instructing was stressed at both sites by the administrators. Both schools had detailed written rules to guide lesson planning. For example, all teachers younger than 35 were required to compose full lesson plans (*xiang an*; *xiang*=detailed, *an*=plan). As specified, lesson plans should be finished one week ahead. The plan should explicate the content of the lesson, standards-aligned teaching goals, important points (*zhong dian*; *zhong*=important, *dian*=point), difficulties (*nan dian*; *nan*=difficult, *dian*=point), using instructional/technological aids, time allocation, instructional procedures, reform-minded pedagogy, in-class exercises and homework, the layout of blackboard writing, and post-lesson reflection. The T.G. Offices at both schools implemented several school-wide inspections during the spring semester. They collected all of the teachers' lesson planning notebooks and gave scores. We photocopied the plans of several focal teachers, which were neatly written with thoughtful post-lesson reflections. All appeared wonderful on the surface review, until one day we found Teacher Zhang (Merits School) was copying plans from a commercial publication. On the right margin of her lesson plan notebook was written in red the post-lesson reflection, though she had not taught the lesson yet. Afterwards, another teacher in the office copied from Teacher Zhang's notebook. "Don't take those plans too seriously," Teacher Liu (Pioneer School) laughed. "We rarely prepare a lesson that way – that is to meet the requirements of the school. It is what is in our mind that matters, not on paper!" (Teacher Liu, Informal Conversation in Grade 3 Group).

In this study, we observed lessons, school meetings, teacher professional development, and teachers' everyday work in offices in a 7-week span. We spent more than 150 hours in the teachers' offices at each school. We took field notes when we were observing, which amounted to more than 120 pages. Specially, we observed 12 lessons at Merits School and 11 at Pioneer School, and they were video recorded. At both schools, we attended teachers' weekly teaching research activities and weekly school-wide meetings for seven weeks. We also took part in within- and out-of-school professional development. For instance, we participated in the Grade 6 lesson observation when one inexperienced teacher gave an open lesson at Merits School. Another example was we joined Grade 1 teachers in the 3-hour district-wide Textbook Analysis and Training at the beginning of the fall semester.

Document review. Another source of data we drew on in this research was from over 1,500 pages of school records and archives, policy documents, and student and teacher artifacts. Objectively speaking, documentary data have several advantages over interview and observation. First, documents tend to record events over a long period of time. Equipped with those documents, we managed to trace back in time the original decisions made, people involved, actions taken, and outcomes yielded in implementing the N.M.C. at both sites. Second, documentary data may help offset the flaws in human memory that might happen in interviews. Moreover, artifacts are more tangible and vivid products that can be used to corroborate or test narrative data collected via interview and observation.

In our cases, documents and artifacts had to be used judiciously, however. The preceding vignette actually points to the tricky issue, that is, how to treat documentary data. Clearly, these materials were not designed for the purpose of research and represented the vested interests of the schools and teachers. The strong subjective and utilitarian propensity could bias records. We adopted a discerning threshold in amassing data and the subsequent analysis. Instead of treating documents at face value, we always consulted with the administrators and teachers to find out if a particular record was authentic and trustworthy.

The first kind of documentary data purported to reconstruct the schools' histories and situate them in the ongoing flow of time. Archives at Merits School were well maintained, some dating even back to 1978. The archives covered a wide range of topics, even including, for example, enforcing the one child family policy in the school. We reproduced its Annual School Plan and the Yearly Report for the years from 1982 to 2018. In those materials, the major achievements over the past year in the school and what to pursue in the coming year were generally documented. Often, monthly highlights were chronicled on a separate sheet. For instance, Table 3.6 displays the contents of the monthly highlights for Merits School during a portion of the school year.

Documents, such as the school policy collection, regulations on teaching norms, and national and school curriculum schedules were garnered too.

Another source of documents came from teachers' and students' work. We gathered teachers' lesson plans, teaching materials made by the teachers, and diaries related to teaching. Also, we obtained students' work, including homework, commercial workbooks, exercise books, and examination papers. School-based *teaching research* was a key form of professional development embedded in the workplace. We reprinted the T.R. records from the spring semester. These materials documented major teaching research activities. The last batch of documentation concerned school administrative records. As a way of evaluating teachers, the school frequently inspected teaching norms. We selected and copied the inspection records from March to July.

In a similar vein, we collected archives, from the year 2000 onwards, policy documents, school regulations, teachers' lesson plans, student work and the like at Pioneer School. In particular, the school compiled a book of their experiences with the N.M.C. reform. We made a copy of the book. It was an important source of original records to reconstruct the school's reform history.

Data management

All formal interviews were recorded digitally and transferred into a computer. We transcribed them and stored interviews from different participants in separate document folders. Informal conversations with administrators or teachers in the playground or in offices were recalled from memory and typed

Table 3.6 Monthly highlights in the school year at Merits School

August
(1) The district organized experimental teachers for training on the new curriculum
(2) City-level backbone teachers attended the city training.

September
(1) National Teacher's Day celebrated on September 10.
(2) District T.R. Center held district-wide lesson observations at our school.
(3) School administrators conducted random school-wide lesson observations.

October
(1) The school obtained first place in the 10 kilometer cross-country race.
(2) One teacher went to the provincial capital for the Standard Mandarin Training.
(3) The school obtained the Key School of Basic Education Award.

November
(1) Chinese teachers visited cooperative elementary schools in Beijing on November 16.
(2) Mathematics teachers visited cooperative elementary schools in Beijing on November 19.
(3) The District Technology-aided Education Office came for an inspection on November 27.
(4) The school T.G. Office inspected students' homework and teacher's teaching plans.
(5) The Principal conducted school-wide random lesson observations.

December
(1) Tested teachers on the ideas and theories of the N.M.C. reform.
(2) The school held an open house for parents.

January
(1) The Bureau inspected the school performance.
(2) Review and preparation by the school for the final exam.

up on the computer later that day. Twenty-three videos were also transcribed and generated about 160 pages of data. Field notes were handwritten and kept away from the reach of teachers. Together with 43 memos, they were typed up on the computer at home later. We sorted all documents and classified them into 24 different categories. The file folders were stored securely in a file cabinet. All electronic data were backed up on another computer and a hard drive.

All original data were recorded in Chinese. When citing data excerpts in this book, we translated them. During the process, we turned to colleagues who were good at both Chinese and English to scrutinize and crosscheck the translation. Sometimes, some features and meanings unique to the Chinese language were reduced and even lost in translation. In such instances, word-by-word explanations of particular terms or sayings are offered in order to facilitate the readers' understanding.

Data analysis

The essential feature of the case study methodology is that data collection and data analysis take place simultaneously and both processes inform each other interactively to culminate into a refined descriptive theory or explanation of the phenomenon. Yin (2003) points out that there are four tenets that characterize high quality analysis of case study data: (1) Attending to all the evidence; (2) addressing all major rival interpretations; (3) addressing the most significant aspect of the case study; and (4) utilizing the researcher's prior expert knowledge. Despite these general principles, Yin (2003) maintains that the case study methodology lacks specific strategies to approach data.

In this study, we borrowed some mature techniques in qualitative research so as to strengthen the viability and power of case study data analysis. Huberman and Miles (1983) provide an elaborated procedure for data gathering and analysis, which consists of *coding* (generating categories), *policing* (detecting personal bias), *dictating field notes*, *connoisseurship* (knowledge of issues and context), *progressive focusing and funneling* (narrowing data as the study progresses), *interim site summaries* (summarizing preliminary findings and identifying questions not addressed sufficiently), *memoing* (writing down emerging issues), and *outlining* (developing a standardized writing format for cases). For instance, *open coding* of data was useful to bring themes to the surface from deep inside the data; *selective coding* would be the subsequent step to crystallize broad categories into more abstract, general, and analytically incisive themes (Charmaz & Belgrave, 2001).

We employed iterative inductive processes of gathering, coding, and categorizing data that called for constant comparisons of data bits, codes, and categories to refine categorization. In earlier stages, data at hand were read through, knowledge from literature was drawn on, and flashes of insight were noted. We generated basic codes that were collated into different categories. Then, recurring categories suggested preliminary themes. As more data were accumulated, a new round of coding and theming was carried out. New data might confirm, contradict, or expand previous categories and themes. This process continued until the theorizing was complete. In so doing, the picture of the process and issues in question could be inductively built up. A general principle we abided by in theming was that a theme was invoked more due to its significance than to its frequency. Not all themes are reported on in the book.

We intended to use phrases that the informants said, or actions they committed, as basic codes, since they generally were more vivid and had catchy handles. For instance, Teacher Yan, Grade 3 teacher at Pioneer School, pointed to herself and another colleague, Teacher Hua, and said to me, "One cow! Two cows![6] Grading students' homework the whole morning even without raising our heads!" (Observation). They had been correcting students' work for about 2.5 hours (from 9:00 am to 11:30 am). We then picked "one cow and two cows" as the code to depict teachers' experience with homework.

Not all codes allude to the same dimension or connotation of information. Other codes like "grading face-to-face," "grading in class," and "grading by peers" also concern homework, but they are more on the varied methods to grade homework. The code "one cow and two cows" also reflects the teacher's negative emotional reaction to the overburdening workload. The above codes could be categorized as "experiencing homework." By grouping like with like, categories of "experiencing homework," "preparing the lesson," "instructing" and the like culminated into the theme, "teachers' work: Onstage and offstage."

Spending a longer time in the field proved to be an effective way to enhance the veracity of our findings. At Merits School, Teacher Hong's case was heuristic. When we first met her in late June, she sounded very upset and outraged, "Reform! Reform! Why the more they reform, the worse it gets?! China's education is completely hopeless!" (Teacher Hong, first contact). Her words immediately set a gloomy tone regarding the outcome of the curriculum reform. We concluded that the curriculum reform was apparently a failed attempt. In the following fall semester, when we started observing her classroom instruction, however, we were surprised at her constant use of small groups. From time to time, she asked students to discuss with peers in a pair of two or group of four. Admittedly, the small group method was still applied in a rather rudimentary fashion, but it demonstrated a fundamental shift from the traditional "stuffing the duck" pedagogy[7] that we had expected to see. She was not putting on a show for us either. Several times we conducted the observation outside her classroom without her knowledge, and found her frequently providing students opportunities to work with peers. Seeing that, we had to challenge the prior theming. At least, her behaviors seemed to be revised. The follow-up interviews with her confirmed that the reform had resulted in substantive change in her beliefs, knowledge, as well as instructional practices. The reason that she had been totally negative about the reform during our first conversation was because she felt too contrite about having to oppress students to drill for the district Uniform Examination. The school's emphasis on the Examination in conjunction with a set of other factors precluded her from fully actualizing reform ideas. Hence, we revised the prior theme and termed it "bounded change."

Each case study was written in a narrative format respectively from the perspective of administrators, teachers, and parents. Their experience with and understanding of the curriculum reform and education in general was described at greater length in order to "establish an empathetic understanding for the reader, through description, sometimes thick description, conveying to the reader what the experience itself would convey" (Stake, 1995, p.39).

Ethical issues

The power of the researcher

Our powers as researchers was exercised in multiple ways. Firstly, we were the instrument with thoughts, feelings, and judgment, subject to interpersonal chemistry in data collection. We had a large degree of liberty to decide what to look at and what not to look at. However reflective and unbiased we attempted to be, we might still selectively pay attention to phenomena that apparently raised our awareness or piqued our interest. Our familiarity with and knowledge of Chinese culture might be a barrier to us. We might have failed to recognize the uniqueness of certain educational phenomena, because they seemed too commonplace to me. To overcome this limitation as much as possible, we adopted a critical lens to constantly question ourselves and to reexamine the familiar.

Secondly, the phenomenon was filtered through our interpretation. In data analysis, we kept alert to the danger of misinterpretation. In this case, the technique of member checking was used to verify the data and tentative interpretations. If they were available and willing, we gave the interviewees the transcripts for proofreading and gleaned their feedback on our preliminary analysis for authenticity of representation. We were thereby made aware of any bias and avoided misinterpretations to a greater extent. Member checking also helped ground our emerging findings and explanations in the local sociocultural context.

Lastly, our pen had power. One teacher we observed said to me, "Don't depict the Chinese education that backward. How much it will lose face?" She was half joking and half serious when she said that. Her words illustrated the power that we as the researchers had. No matter how fluid and vague the phenomenon was, it would become extremely daunting once we put it down in words. The written product became the sole representation of the multiplicities of the reality. Should we hide something in order to make our fellow Chinese teachers, and ourselves, less embarrassed? Should we manage the delicate balance of what to tell and what not to tell in presenting this work? On the one hand, we could paint a perfect picture of Chinese mathematics education if we did not regard the broader sociocultural context it was situated in: Look, how profoundly China's mathematics teachers master content knowledge, what a sophisticated system of professional development for teachers they have, and how impressive the students' test grades were. On the other hand, we needed to curb the propensity to only expose the "dark side" in the hope of courting the curiosity of readers, since such a dark side exists for a reason. Even though it was frequent for us to feel sad, angry, or depressed throughout the process of the study, we managed to distance ourselves from our emotions and look at the same phenomenon from different angles, as different informants made sense of them.

Confidentiality and privacy

Protecting the participants from potential repercussions is our first priority. School administrators and teachers are at the bottom of the educational hierarchy. We left the City after the study was concluded, but they still work there, subject to potential risks from powerful higher-ups. This is a small world, particularly so in this City. To keep the data confidential, we kept our notes away from others while in the field. We did not discuss issues related to one site with people at the other site. If we publish this work in Chinese in the future, readers could still easily identify where and who we are talking about. Therefore, it is necessary to take some protective measures in reporting data. The exact location of the City was hidden, the gender of certain informants was not disclosed, and stories of one person were broken apart and mingled with others.

Travelers on a train or airplane are more prone to open their heart to fellow travelers. When faced with strangers, people shed more easily the masks that they tend to wear when interacting with acquaintances, probably because it is unlikely they will ever see the stranger again. Perhaps it was because we were from outside of the school that teachers confided in us. They requested confidentiality. It was a request that could not be refused. We owed the teachers a moral obligation to keep their stories anonymous or keep them to ourselves to protect their identities, even though they could have potentially led to a very important research theme.

A comparative lens

Having a certain amount of first-hand experience in both China and America and straddling two different cultures, we look at ourselves more as academic brokers more as an academic broker and view the study as an opportunity to engage ourselves in an academic exchange. As people who grew up in Chinese society, we are knowledgeable, to a certain degree, of how Chinese schools function and what issues exist. Also, life in the U.S. has exposed us to a different culture and merged a new perspective in my worldview. As a result, we tend to view things through a comparative lens. When making sense of a phenomenon, we are inclined to decipher it by summoning our knowledge of both cultures and recount it with the audiences of both sides in mind. Our writing was a vehicle to achieve these ends.

Notes

1 China's government system consists of the central government and the local government. The latter contains, in order, *provinces*, *cities*, *counties (or districts)*, and *townships*. *County* and *district* are interchangeable in this manuscript. The term *district* should not be interpreted as *school district*.

2 For the sake of confidentiality, we do not give the exact G.D.P. ranking of the City in case the readers could easily locate the place through precise information. The statistics are from online sources.

3 Teaching norms are a teaching inspection and evaluation system institutionalized in schools across China. With some variations, teaching norms refer to five key teaching subprocesses: Lesson planning (*bei*; *bei*=prepare), instructing (*jiao*; *jiao*=teach), student work (*zuo ye*; *zuo ye*=work), individualized tutoring (*fu*; *fu*=counsel), and testing or assessment (*kao*; *kao*=test). The norms establish the standards to which teachers should conform in teaching activities and specify how teachers should be monitored and appraised.

4 The official level of backbone teachers corresponds to that of the government structure. Respectively, there are county/district-level, city-level, province-level, and nation-level backbone teachers. Teachers earn these honor titles through district-, city-, and province-wide lesson competitions in corresponding subject areas. Individual schools often identify their own backbone teachers.

5 All names of school employees in this book have been changed for confidentiality reasons.

6 In Chinese, "cow" refers to being hard working without making any complaints.

7 "Stuffing the duck" or "pouring food into the stomach of the duck by force", are expressions in Chinese to describe students being fed with knowledge by rote learning.

4 Merits School confronted with the reform

The initiation of the reform

The reform at Merits School

The curriculum reform, aiming at reworking all school subjects, was officially unveiled in the Red Pebble District in early 2002. V.P. Yu of Merits School, also a former mathematics teacher, noted that the reform initiatives that had been receiving national attention for the past 3 years were "eventually coming their way" (V.P. Yu). Amid a series of high-level proclamations from the State Council (1999) and the M.O.E., the New Curriculum Reform left T.D. Wang, a mathematics teacher in Grade 6, with a deep impression. In fact, in October 2001, as one of the 38 National Experiment Districts for the New Curriculum Reform, a city in an adjacent county had been designated and started testing the waters. Soon after that, the order to launch the Reform was passed down level by level, nationally. In the province, a Provincial Steering Committee, including the head, vice heads of the Province Education Bureau, and several high-ranking officials from the provincial administration, had been appointed for the overarching orchestration and oversight of the province-wide implementation of the curriculum change. According to the national Action Plan (M.O.E., 2001), every city in each province should establish at least one province-level experimental zone (which overlapped with the administrative county or district). Together with another 15 counties (out of 91) across the province, Red Pebble was singled out as the only Provincial Experimental District in the City.

Following closely in the province's footsteps, the City formed its city-level committee in early March 2002, which was made up of heads of the C.E.B., the director of the City T.R. Office, and heads of D.E.B.s. In a similar fashion, the Red Pebble District set up its steering committee too, including a dozen officials from the District Legislature, the Education Bureau, and the District Teaching, Research and Training Center (for short, the District T.R. Center)[1].

On May 12, 2002, the reform was officially launched in the City. That day, the city-level steering committee held a conference and announced its formal commitment to the reform. The meeting intended to boost educational officials and school leaders and resolved to mobilize educational personnel and resources in the City to undertake the reform.

In the following week, the D.E.B. convened elementary and middle school principals across the district to further deploy the reform. Principal Li, the then principal, and V.P. Yu were both required to attend that meeting. During the meeting, in reference to the national policies, the rationale for and the imperative for implementing the reform were once again elaborated on. Enthusiastically, the D.E.B. called for "a scientific, systematic, and immaculate approach" (Red Pebble D.E.B., 2002, p.1) to full implementation of the reform. Executive power was entrusted to the District T.R. Center. The District T.R. Center would lead, coordinate, and oversee schools to carry out the reform step-by-step.

Even though V.P. Yu had taken part in numerous educational reforms, large and small, lasting and short-lived, in her 25-year teaching career as a mathematics teacher and a school administrator, the approaching reform had projected far more ambitious goals. First, the scope of the reform was unprecedented. The reform focused on all grade levels and all school subjects. Second, the management of curricula would be decentralized. Instead of having one single national curriculum like before, provinces and schools were allowed to offer their own local courses. Third, the whole gamut of curricula would be redesigned in line with one coherent system of objectives, the three-dimensional teaching objectives, that is, (mastering) basic knowledge, (mastering) basic skills, and (attending to students') affection, attitudes and values. Deriving from this system, the mathematics curriculum reform focused on four objectives: Knowledge and skills, mathematical thinking, problem-solving, and affection and attitudes toward mathematics. Lastly, revolutionary pedagogy was proposed. For the first time, she was informed of small group, inquiry-based, cooperative learning. The reform's massive scale, depth, and comprehensiveness all led V.P. Yu to think that the state was determined this time to overhaul the current educational system and to make a real difference in the new millennium.

Merits School committee

Being one of the high-performing schools in the district, Merits School was very much expected by the District T.R. Center to play an exemplary role for other schools to learn from. To V.P. Yu, it meant recognition from the district, but also the applying of pressure. Situated at the bottom of the power chain, Merits School was left with no room to negotiate about the reform. "It was not up to our school. The higher-ups made the decision, and we just followed it out," V.P. Yu remarked when expressing her first reaction to the reform (Interview #1).

She was no stranger to reform directives of the D.E.B. or the District T.R. Center. She recounted there had been at least three local attempts at mathematics teaching reform before 2002. As early as 1984, when V.P. Yu just started teaching, the district adopted a new approach to mathematics teaching, designed by a star teacher of mathematics in Beijing, Ms. Xinlan Ma[2]. She emphasized that mathematics teachers should not only teach students basic knowledge and concepts, but enable them to develop all-round intelligence and skills. Phrases such as "alleviating the workload of children," "developing all-round children," "respecting students' ownership of learning," and "teachers are the leading facilitators" were popular. Techniques of improving oral and written computation, and of solving word problems were incorporated to maximize instructional efficiency in a lesson lasting 45-minute.[3]

The experiment lasted until 1993 and was replaced by another emerging focus. Students' heavy academic burden and rigid learning were once again under fire. The Experiment on Modern Elementary Mathematics Teaching was initiated and it had three goals: (1) Letting teachers "completely get rid of the biased mentality of solely pursuing test grades and reestablish positive beliefs about cultivating children's intelligence and developing their abilities" (Merits School, 1993, p.1); (2) abolishing the out-of-date "stuffing-the-duck" (*tian ya fa*; *tian*=fill, *ya*=duck, *fa*=method) teaching method and replace it with the student-centered, inquiry-based approach; and (3) emphasizing teaching research to prevent teachers from the stagnation of "only teaching without researching" (Merits School, 1993, p.2).

Before this experiment was finished, another fad of educational innovation swept across China in 1997. That was promoting *qualities-oriented education* (*suzhi jiaoyu*; *suzhi*=qualities, *jiaoyu*=education), coupled with the cries for *education for creativity* and *education for innovativeness*. The core idea of qualities-oriented education was promoting the all-round development of students. As it used to do, the district shifted gear mid-course and directed Merits School to follow the emerging national directives. The qualities-oriented education movement culminated in the standards-based curriculum reform, which took shape in the early 2000s and intended to address the aforementioned learning issues that held on stubbornly for the previous two decades.

As a response to the coming reform in 2002, the first step Merits School took was to set up a steering committee to demonstrate its affirmative reaction. Mirroring the district's managerial model, Merits School set up its own leading team to orchestrate this reform. Principal Li assumed the chair position, with V.P. Yu and another V.P. as the deputy chairs, and the school party secretary, the head of school labor union, and T.D. Wang as members. In fact, all levels of administrator appeared on the list, as a gesture to show that the school leadership was unanimous to carry out the reform. As usual, no "ordinary" teachers or parents sat on the committee. They had little input with regard to what was going to happen at the school. Their opinions were not even solicited. One mother said:

> We don't know much about the reform. The school did not talk to us about that. We only know the curriculum is getting tougher. When my

> son brought home some problems, we weren't even able to solve them... In high school, mathematics was my best subject. But, we did not dare to tutor my son at home.
>
> (Mother 1, Interview #1)

Another parent concurred. She stressed that they looked up to the teachers to make sure their children had good grades. Beyond "the curriculum is getting tougher" (Mother 2, Interview #1), parents could not tell much about the N.M.C. They had not been informed by the school of the meaning of the NMC and how it would be different.

V.P. Yu and T.D. Wang were the ones who were actually invested in the day-to-day details of the reform. V.P. Yu faced a more delicate situation, positioned in between the district and the school, and between school administrators and teachers. On the one hand, she had to be responsive to the district's calls; on the other hand, she needed to gain support from the Principal, T.D. Wang, and the majority of the teachers. The latter, however, were specifically held accountable to assuring the quality of teaching of the school, which constituted her primary responsibility and concern.

Following this move, V.P. Yu and T.D. Wang formulated a number of school plans, including the 6-year Reform Plan, Procedures for the Experiment, the Teacher Training Plan, and the Incentive Plan for Teachers with Outstanding Reform Achievements. In the 6-year Reform Plan, V.P. Yu detailed the steps to take and the goals to achieve after the 6-year cycle of reform. The school placed a primary emphasis on classroom teachers, in particular early career teachers, and claimed that "the reform's success depends on teachers, on whether they can rapidly change their attitudes to teaching" (Merits School, 2002a, p.2). Thus, among the six issues discussed in the Plan, three were about classroom teachers: Updating teachers' knowledge of the reform theories, strengthening teacher professional development, and promoting teaching research (*jiao yan*; *jiao*=teaching, *yan*=study or research). First, to help teachers learn up-to-date educational theories, the Plan stated that in the beginning of the coming fall semester (in September 2002), they would organize teachers to study collectively official guides on the N.M.C. standards. Meanwhile, they would periodically disseminate learning materials and hold special workshops. Teachers would need to regularly turn in written reflections and papers on theory learning. Second, the Plan proposed that the school would pay close attention to teacher professional development throughout the implementation of the reform to transform teachers' attitudes and instructional behaviors. Third, the school would strengthen teaching research activities so as to inform teachers' teaching through research. For these purposes, a hierarchical teaching research network made up of school principals, the T.G. Office, and grade-level T.R. groups would be established. Centering around the network, the school would conduct regular teaching research activities ranging from collective lesson planning, to lesson observations, to teaching competitions and the like.

These plans that V.P. Yu and T.D. Wang drafted often contained dressed-up information. For instance, the T.R. system was not some new innovation, but had already been established nationwide in the late 1970s. It constituted an essential element of Merits School's teaching norms. Indeed, such information was commonplace in every one of the school's reports to the D.E.B. From 1993 to 2002, Merits School repeatedly reported that it had such statements as "stipulated firm measures to decrease students' workload," "effectively reduced the amount of and refined the types of homework," and "diversified [approaches to] assessment of student learning". In reality, its examination-oriented status quo did not get changed.

The aforementioned plans had never been referred back to since their stipulation, however. But the school had to work to compose those plans, as V.P. Yu pointed out, since the higher-ups (*shang mian*; *shang*=up, *mian*=face) would come to inspect the school and "they are particularly attentive to the paperwork" (V.P. Yu, Interview #1). An inseparable part of V.P. Yu's job as the academic V.P. was to prepare written materials. It was an important component on occasions when outside officials came to inspect the school, or visitors came to learn the school's exemplary practices. Merits School would lay the documents out in front of the inspectors and visitors as examples of the school's achievements and efforts.

For administrators at Merits School, actions like forming a grandiose but mostly inert steering committee, or making impressive plans was more about making a symbolic, rather than a substantive, gesture. It demonstrated to the higher authorities that they were carrying out the mandates conscientiously. By doing so ceremonially, Merits School claimed and maintained its legitimacy in the face of the ever-changing external policy environment, which left room for the school to navigate in the environment to its own advantage.

Training in the province

Merits School's specific reform preparation began in the summer of 2002. A 40-day, 260-hour training event at the province-level about the new curricula was held at another city during the summer vacation. Nearly 1,000 people attended the training event, including elementary school principals, T.R. fellows, and province-level backbone teachers across tens of subject areas from the 16 experimental counties. The Red Pebble District dispatched about 100 T.R. fellows and principals too. V.P. Yu was the only teacher selected from Merits School to attend the event. She was both the academic 6 V.P.s and a province-level backbone mathematics teacher, and was supposed to conduct teacher training back at her home school.

The training event purported to enable the attendees to "understand and grasp the contents and pedagogy concerning the new national curricula and be able to competently play the role of a backbone back to their regions" (Province Teacher Training Center, n.d.). Six modules of coursework were offered, and a number of topics were covered, for instance, the ethics of teaching, interpretations of curriculum standards, the analysis of textbooks, and modern instructional technologies, just to name a few. In particular, training focused on three aspects: Reviewing the background of the new reform, interpreting new curriculum standards, and analyzing textbooks. A few professors from one university in Beijing were invited. Those scholars had been involved in drafting the N.M.C. standards and composing standards-based textbooks. They were among the most authoritative people in this field in the country.

V.P. Yu rated herself as very "receptive to" the ideas promoted during the training, because she also considered that it was time to change. As a grassroots teacher, she was not oblivious to the gnawing issues deeply rooted in China's mathematics education, and more broadly, basic education. "Students are too burdened because of the overemphasis on tests," V.P. Yu lamented. "Their potentials are dried out" (Interview #1). The training event reassured her that the higher authorities were very likely to make a real change this time.

"To be able to lead teachers to pursue the reform," V.P. Yu went on to state, "we must keep abreast of the new thinking and proactively update our knowledge and attitudes" (Interview #1). In terms of enhancing her understanding of the reform ideas, the training event served her well. But most training during the 40-day period of time was conducted in the form of expert-centered lectures. The fact that "you sat there listening to the experts" (V.P. Yu, Interview #1) did not help her much to visualize what a reform pedagogy should look like.

She was not the only one who experienced the inability to translate those theories into actual instructional practices in classrooms. Back home, Principal Li and T.D. Wang, together with teachers in different subject areas at Merits School, attended a one-day workshop organized by the D.E.B. Similarly, general information about the curriculum reform and its necessity was conveyed to them by professors invited from outside. The training event was not engaging, as T.D. Wang recalled; it was too theoretical and abstract. What was lacking at that time was that no schools in the district had really used the N.M.C. Without having a concrete image to turn to, V.P. Yu and T.D. Wang felt that they were fumbling in the dark. Teacher Wu, then a Grade 1 mathematics teacher, observed the disorientation of the school administrators who were supposed to guide her:

> We remember that the reform in the beginning was very boisterous, very sensational, the whole school, the whole district. What we used to do was all of a sudden invalid. The school [leaders] asked us to renovate our teaching. But, they only gave us an elusive orientation. In terms of how to get there from here, we were left on our own to figure it out. V.P. Yu hasn't taught for years. T.D. Wang herself is a teacher like us. We felt we were all lost – we did not know what was the right way to teach…even what to teach and to what degree. That kind of feeling.
>
> (Teacher Wu, Interview #3)

Due to the superficial training they received and the limits of their own experience and expertise, neither V.P. Yu nor T.D. Wang could provide teachers with the necessary modeling. They had undergone the same system of education as other teachers and practiced accordingly for many years in the past. If others were teaching in the "stuffing-the-duck" approach, theirs was not any better. In other words, they could not competently play the role of reform leaders. Besides, they were not sure what was in the mind of the district T.R. fellows and how they envisioned the reform. After all, those higher-ups had the final say.

Reform in action: The stage of authentic implementation

Vice principal Yu: The reformer

At first, V.P. Yu was determined to execute the reform to its full extent. In September 2002, Merits School adopted a range of new curricula. The new mathematics curriculum concurrent with other subjects was implemented on the incoming Grade 1. Other grade levels kept using their old materials until the whole series of curriculum were completed. That year, China extended the length of elementary education from 5 to 6 years; thus, the incoming Grade 1 in Merits School had 12 classes, half on the 5-year track, and half on the 6-year track. The N.M.C. materials, developed by one southern province in China and more progressive and challenging, were adopted only among six classes on the 6-year track, while the other six classes still used the more traditional curriculum that was also slightly modified according to the N.M.C. Standards (2001). The six teachers teaching the 6-year track were appointed as experimental teachers to participate in the district and the school's professional development activities. Teachers on the 5-year track, as well as mathematics teachers in all other grade levels, were required to observe and learn from experimental teachers, and apply new ideas in their own teaching.

In the fall of 2002, the Weekly Curriculum and Class Schedule formulated by the Province Education Bureau was put into effect for all grade levels in Merits School. According to the Schedule, Grade 1 and Grade 2 should have 25 lesson periods per week, including four 40-minute periods of mathematics, eight of Chinese, four of physical education, four of arts, two of moral education, one of reading, one of handcraft, and one school-based subject; from Grade 3 onward, students should have a total of 30 periods of lessons per week, and science and English were required. As for mathematics, there should be four lesson periods of mathematics for Grades 1–3 and five periods for Grades 4–6 weekly.

In the meantime, echoing M.O.E.'s policy to reform the system of assessment and examination (M.O.E., 2002), the District T.R. Center announced that they would put a stop to the district-wide Uniform Examination (*tong kao*; *tong*=uniform, *kao*=test). Relying on test grades, the sole yardstick to assess performance of students and schools had long been denounced as the pathology of China's education. The District T.R. Center tried to re-direct its focus on developmental assessment throughout the whole learning process. The Examination used to be administered at the end of school year. Previously, all schools in the district would be ranked against one another by their average grades in the Examination and the ranking would be broadcast afterwards to make all schools conscious of their own standing.

Thus, "it all boiled down to one thing – the test score," V.P. Yu remarked, "It was the only thing that the District T.R. Center used to evaluate a school's educational quality" (Interview #1). Top ranking would squarely attest to the excellence of a school's leadership, and vice versa. Over its long history, Merits School had performed superbly at all grade levels in contrast to its peers in the district. Rarely had it slipped to second place, which had made V.P. Yu very proud and concerned. Pressed by the district's testing and ranking policy, she alleged, Merits School had been left with few choices but to focus on preparing for the Examination. As a result, the Examination had been the most critical issue confronting the school leaders in the past. Now that the District T.R. Center would give up that baton and engage in actual change, V.P. Yu became motivated to carry out the new reform authentically.

V.P. Yu attempted to tackle the issue through: (1) Enforcing the official curriculum schedule, and (2) strengthening reform-aligned teaching norms to change teachers' practices. Once again she put forward several regulations. This time, however, it was different from the past: Not only did she stipulate rules, but she actually enforced them.

Enforcing the official curriculum schedule

The first measure was that V.P. Yu required teachers to faithfully conform to the provincial Weekly Curriculum and Class Schedule. V.P. Yu admitted that the total mathematics hours per week given in the official Schedule were too few, so she added one more period to every grade level. In comparison with Merits School's previous curriculum plan, the total number of mathematics lessons per week was significantly reduced from an average of 11 to five. Teachers were not allowed to teach mathematics during the hours allocated to physical education and art.

Take Teacher Wu's Grade 1 curriculum schedule as an example; it is displayed in Table 4.1. A full range of subject areas are shown on it. Note in particular: The total number of mathematics lessons per week was reduced to four periods. (Since 2006, the total number of mathematics lessons for Grade 1 had been increased to 14 periods.) The only difference from the Province Schedule was that it had 33 periods of lessons per week instead of 25, or roughly seven periods per day. On Tuesday and Thursday afternoons, students were dismissed one period early, since teachers needed to use the last lesson to conduct teaching research.

Reducing the number of lesson periods was sharply against the traditional practices of mathematics teaching at Merits School. On average, they used

Table 4.1 Teacher Wu's Grade 1 curriculum schedule at Merits School

	Mon	*Tue*	*Wed*	*Thurs*	*Fri*	*Sat*	*Sun*	*Notes*
1	Math	Chinese	Math	Chinese	C.C.			
2	Chinese	Math	Chinese	Math	Chinese			
3	Break	Break	Break	Break	Break			
4	Blank	Art	Music	Blank	Chinese			
5	P.E.	Chinese	Moral	P.E.	S.B.C.			
6	Blank	Moral	Library	Art	Music			
7	Chinese	P.E.	Handcraft	S.B.C.**	P.E.			
8	Break	off	Break	off	Break			
9	C.M.	off	P.E.	off	Culture			

Note: C.C.=creativity cultivation, S.B.C.=school-based curriculum, C.M.=class meeting.
*Crt=creativity cultivation; **S.B.C.=school-based curriculum; ***C.M.=class meeting.
All periods highlighted in the table were used for mathematics

to have at least two periods of mathematics per day: The first period was to teach new knowledge, and the second was to do exercises to consolidate the content learned in the first lesson. To enforce the official Schedule, V.P. Yu demanded that T.D. Wang and her T.G. Office inspect frequently whether teachers were conforming to the Schedule. Thus, T.D. Wang and her associate directors rotated to observe teachers' classes at random and to make sure that they were implementing the plan.

Reforming teaching norms

The second measure that V.P. Yu together with T.D. Wang and the T.G. Office took was to try to incorporate reform ideas into extant teaching norms and to stipulate new rules about lesson planning, instructing, student work, tutoring, and assessment. Specifically, lesson planning should embody three-dimensional objectives (that is, (mastering) basic knowledge, (mastering) basic skills, and (attending to students') affection, attitudes and values), arrange cooperative learning, and use manipulatives; instructing should materialize the reform-minded pedagogy; student work should be stratified and diversified so as to cater to different ability levels of students; individualized tutoring should enable students to develop differently; and assessment should be plural, attend to the learning process, and aim at motivating students rather than dampening their interests.

Lesson planning. The curriculum reform lent new ideas to the school's norm on lesson planning. In accordance with the reform, lesson planning should be "innovative and dare to smash the limits of dated notions" (Merits School, 2002, p.1). The focus on which V.P. Yu and the T.G. Office placed was that lesson plans should spell out how to realize the three-dimensional objectives, a key emblem of the N.M.C. Reformed teaching methods, for instance, cooperative learning, received particular attention. Teachers were asked to

Table 4.2 The progress guideline for Grade 1, 2nd semester

Week and date		Instructional content	Pages
1	03.02–06	A – position.	1–9
2	09–13	B – subtraction within 20 by decomposing.	10–14
3	16–20		15–18
4	23–27		19–26
5	03.30–04.03	C – shapes and figures.	27–30
6	06–10	D – knowing numbers up to 100.	31–37
7	13–17		38–45
8	20–24	E – using money.	46–55
9	27–31	F – addition and subtraction up to 100.	56–66
10	05.04–08		67–73
11	11–15		74–80
12	18–22	G – telling the time.	81–87
13	25–29	H – looking for patterns.	88–92
14	06.01–05	I – statistics.	93–97
15	08–12	J – general review.	98–101
16	15–19		102–105
17	22–26	Comprehensive review.	

encourage students to bring objects from home, to conduct observations, or to look up information on the internet.

As before, V.P. Yu asked teachers to prepare three different plans: The term plan, the unit plan, and the lesson plan. The term plan was the teaching blueprint for the whole semester. Teachers should compose the plan before the semester started or within one week of the beginning of the semester. It must be made on the basis of the curriculum standards, the textbook, and the students. It was considered especially important to study and know the students well. The plan should specify the objectives, time allocation, and the strategies to improve teaching quality. Time allocation should refer to the Progress Guideline developed by the District T.R. Center. Table 4.2 gives an example of the progress guideline that all Grade 1 teachers in the district should follow.

Preparing the unit plan was intended to encourage teachers to study the whole unit globally before instructing. It should be briefly written and enriched through individual lesson plans. Generally, the school expected teachers to plan lessons one week ahead. In greater detail, lesson plans described the subject matter, important points (*zhong dian*; *zhong*=important, *dian*=point) and difficulties (*nan dian*; *nan*=difficult, *dian*=point), the type of lesson (one of the four types: New content, unit review, general review, or exercise lesson), the process, instructional methods, use of manipulatives, time, blackboard layout, exercise problems, and post-lesson reflections. In order to prepare lessons well, teachers were strongly recommended to know the whole curriculum from Grade 1 to Grade 6 so that they were clear about the interrelationships of different content areas.

V.P. Yu summarized that lesson planning should have "six preparations," "five points," and "four 'carefullys'" (Merits School, 2000, p.2). Six preparations pertained to preparing the Standards, preparing the textbook, preparing the students, preparing instructional methods, preparing learning strategies, and preparing student work. The five points were thoroughly grasping key points, difficult points, knowledge/conceptual points, skill points, and affective points. The four 'carefullys' referred to carefully studying the textbook and the N.M.C. Standards (2001), carefully designing instructional activities, carefully developing assessment, and carefully planning the organization of writing on the blackboard.

Instructing. To dismantle the traditional teacher-directed mode of instruction and reestablish a new, equal relation between teachers and students, the reform redefined the role of teachers as organizers, guides, and facilitators of student learning rather than masters of student learning. Using small group, inquiry-based, cooperative learning in instructing was where the district started from. Following enthusiastically the District T.R. Center's direction, V.P. Yu strongly promoted the independent-cooperative-inquiry instructional model that operated with small discussion groups in class. She proposed four general principles to reform instruction: Giving time back to students as much as possible, letting students use both hands and the brain to practice as much as possible, empowering students to collect and process information as much as possible, and enabling students to as questions and solve problems as much as possible (Merits School, 2002b)[4]. She asked all mathematics teachers to learn to plan lessons that should incorporate "activities guided by teachers," "explorations initiated by students," and "reflections on instruction" (Merits School, 2002b, p.1). In classrooms, it was expected that teachers should truthfully recognize students' ownership of learning, both stimulating students to take the initiative, and being proactive and steering students to engage in thorough thinking and in-depth interactions. After each lesson, teachers were expected to reflect on issues such as the flow of the instructional process, the outcomes, and problems, etc.

To foster reform-based instructional practices, the T.G. Office adopted a comprehensive evaluation scheme, as shown in Table 4.3.

Having eight categories and 36 subcategories, the evaluation scheme looked at four aspects of instruction: Lesson objectives and contents, instructional strategies, outcomes, and teachers' overall performance and skills. In principle, greater emphases were placed on achieving three-dimensional objectives and engaging students in knowledge generation via small group, inquiry-based, cooperative learning. Using the evaluation form, V.P. Yu reported that together with the school T.G. Office she conducted more than 80 observations during the first semester of the reform.

Student work. Student work, including in-class exercises and homework, was considered the most important means to consolidate, assess, and foster student learning. In line with the reform advocates, V.P. Yu added several requirements for student work. First, teachers should not mechanically rely

Table 4.3 The evaluation scheme of instruction at Merits School

Evaluation elements			*Comments*	*Pts*
Objectives and contents (28 pts)	Objectives (12 pts)	(1) Having complete objectives, and encompassing three dimensions: Knowledge and skills, process and methods, and affection, attitude, and values. (2) Meeting the Standards. (3) Being flexible, highlighting key points, and cracking difficult points.		
	Contents (16 pts)	(1) Grasping the structure of the content, reasonably developing the textbook's potentials, and having no errors in or gaps in knowledge. (2) Connecting to students' prior learning experience, and engaging in constructivist learning. (3) Complementing with proper context and materials, and extending learning. (4) Designing the content conducive to cultivating students' abilities of innovation and practices.		
Instructional strategies and methods (35 pts)	Inspiring (9 pts)	(1) Treating students equally and kindly, and creating a democratic and harmonious atmosphere. (2) Evaluating students positively and encouragingly, and igniting students' learning desire. (3) Prompting students creatively and effectively, and encouraging students to think. (4) Using training methods that are scientific, artistic, and rigorous.		
	Self-initiated (10 pts)	(1) Emphasizing self-learning, ensuring time and space for self-study, valuing self-evaluation, and encouraging the development of self-learning capabilities. (2) Employing properly multiple forms of instruction, e.g. interaction, discussion, operation, and contests, etc. (3) Cultivating a culture of mutual learning and assistance.		
	Inquiry-based (10 pts)	(1) Encouraging students to dare to doubt, raise questions, explore and solve problems. (2) Persuading students to think independently and to express unique views. (3) Engaging students in cooperation, exchange, discussion, and inquiry-based learning. (4) Interacting actively and developing mutually. (5) Guiding students to inquire and research in praxis and experiments.		

	I.T. (6 pts)	(1) Utilizing multimedia to engage students' manifold sensory organs. (2) Using appropriate I.T. in teaching. (3) Using I.T. properly.
Instructional effects (22 pts)	State of learning (9 pts)	(1) Students to have a keen interest in, focused attention to, and strong desire for learning. (2) Students to participate in learning thoroughly and actively. (3) Students to take the strong initiative, and have their creativity developed. (4) Students take great interest in practices and experiments, and have serious and positive attitudes.
	Effects of learning (13 pts)	(1) Students have satisfactorily grasped knowledge and skills. (2) Students have learned how to learn, understood the learning process, and learned efficiently. (3) Students have understood the value of the learning content and generated positive interest. (4) Students have been able to consciously transfer knowledge and skills. (5) Students' knowledge and skills have been improved via the teaching and learning activities.
Teacher qualities (15 pts)	Qualities (15 pts)	(1) Passionate and dedicated to nurturing human beings. (2) Witty, flexible, and orderly. (3) Having essential skills (listening, speaking, reading, writing, doing, and applying). (4) Having excellent organizational skills, and being good at affirmative evaluation. (5) Showing a high sense of responsibility and fulfilment.

Note: Translated from Merits School (2002).

on commercial workbooks, but carefully design representative and classic problems in reference to the N.M.C. Standards (2001) and the curriculum. Second, student work should contain a variety of forms of problems and have clear purposes in mind, and teachers should consider assigning an appropriate amount of extracurricular work related to students' lives. Also, she encouraged teachers to differentiate assignments for students according to the level of the student instead of giving the same set of problems to every student. Third, it was strictly forbidden to assign excessive amounts of work to students. She intended to decrease the amount of homework. She specified the maximum homework time per day for each grade level. In accordance with the reform mandates, she required teachers not to assign homework to Grades 1 and 2, to assign no more than 30 minutes of homework to Grades 3 and 4, and no more than 40 minutes to Grades 5 and 6. She insisted that:

> [Teachers] must achieve *Four Have's*: Having flexibility, that is, assigning different problems for students with different abilities; having effectiveness, that is, assigning quality problems and attending to student development; having feedback, that is, garnering and responding to parents' and students' suggestions about homework; having remedy, that is, responding to issues reflected in homework.
>
> (Merits School, 2002b, p.3)

In general, teachers should mark student work immediately. Teachers could mark student work in several ways: (1) Students finish work at home and teachers grade it in the office; (2) students complete work in class and teachers correct it on site so as to identify issues immediately; (3) teachers might only grade the work of target students; and (4) teachers could mark a student's work face-to-face, particularly for low achievers. Teachers should assess the quality of the work, provide written evaluative comments, and date the work at the end. Teachers were not allowed to have students grade each other's work. After grading, teachers should explain all typical errors in the lessons that follow and make sure that students correct all errors before moving on to a new lesson.

To ensure that teachers dealt with student work effectively and conscientiously, the T.G. Office proposed a point system to measure the quality of teachers' work in this respect. Table 4.4 shows the system.

The system took into consideration the quality of problem design, teachers' handling of student work, and teachers' attentiveness to helping students form good habits. It was underscored that students must take homework and exercises seriously. Handwriting must be neat, and tables and figures in student work must be drawn accurately and formally. In particular, students were required to conform strictly to the specific written format[5]. Whether the student followed the format correctly or not was an important indicator of student success.

V.P. Yu was very serious about these requirements. Teacher Rui, who just began teaching Grade 1 mathematics at Merits School in fall 2002, recalled:

Table 4.4 Criteria for inspecting teacher's treatment of student work at Merits School

Focus	*Elements to evaluate*	*Points*
Question design (2 pts)	(1) Purposeful, aligning the Standards and the instructional content.	1
	(2) Not redundant, appropriate quantity, and individualized.	1
Handling (4 pts)	(1) Must grade all work carefully in red ink, highlight all errors, give proper comments in a legible, concise, and objective manner, and note the date. Not highlighting errors: –0.5 pts/occurrence. Leaving problems ungraded: –0.5 pts/occurrence. Wrongly grading problems: –0.5 pts/occurrence. No date when the work is graded: –0.5 pts/occurrence.	2
	(2) Must conscientiously address all typical errors in student work. Leaving common errors unaddressed: –2 pts/occurrence.	2
Student habits (4 pts)	(1) Must have students keep workbooks neat and orderly Untidy (e.g. doodling): –0.5 pts/student. Ripping out pages: –0.5 pts/student.	1
	(2) Must have students conform to the written format (e.g. date of work, activity number, two lines between different assignments). Substandard format: –0.5 pts/student. Illegible handwriting: –0.5 pts/student.	1
	(3) Must have students finish work in a timely and serious manner. Missing assignments: –1 pt/occurrence.	1
	(4) Must have students treat errors seriously and all must be corrected. Not correcting errors: –1 pt/occurrence.	1

Note: The table was translated by the author from Merits School (2002d).

> At that time, the rules on homework were very strict…She [V.P. Yu] would ask T.D. Wang to collect students' workbooks and check if we really did that. She treated it seriously. If we didn't meet the requirements, she would call us to her office to give us a warning.
>
> (Teacher Rui, Interview #2)

V.P. Yu also stressed the importance of diversifying the types of homework given, which was also one of the reform initiatives. Teacher Rui said:

> She [V.P. Yu] didn't like us to assign homework to lower grade levels, like Grade 1 and Grade 2…she didn't want us to give the whole class the

> same problems. We were asked to diversify [forms of] homework. For example, a good student should be provided with something more difficult, something that could elevate his or her current understanding. Average students would be given something less challenging. But a child with a lower level just needs to do the basics, like, he or she could ignore word problems if they were beyond his or her head, and he or she could just work on sums. Just like that.
>
> (Teacher Rui, Interview #2)

Rather than simply having students solve repetitive paper-pencil problems, V.P. Yu encouraged teachers to assign homework that broke with tradition, like collecting life objects at home, making journals, and so on. Even today, Teacher Rui still occasionally asks her students to keep journals about the mathematics they experienced in everyday life.

Tutoring. Tutoring was also one of the teaching norms at Merits School, which was described as an activity inseparable from the handling of student work. Referring to the reform advocates, V.P. Yu stressed that tutoring should be holistic, that is, aiming for all students: For high achievers, teachers should afford them the opportunity to pursue more challenging mathematics; for those in the middle, teachers should turn them into high achievers; and for low achievers, teachers should help them improve. In order to guide teachers to systematically carry out individualized tutoring and to keep track of their work, the school adopted six forms created by the C.E.B. They were, respectively, the Tutoring Plan for High Achievers, the Log of Tutoring High Achievers, the Summary of Teaching High Achievers, the Tutoring Plan for Low Achievers, the Log of Tutoring Low Achievers, and the Summary of Teaching Low-Achievers. At the end of each semester, teachers were required to turn in those forms.

Assessment. Another issue to address that V.P. Yu took seriously was to "get rid of the biased mentality of solely pursuing test grades" (Merits School, 2002b, p.2), the persistent headache that overshadowed mathematics education at Merits School. The District T.R. Center also changed its policy on student assessment. In principle, there should be no Uniform Examinations and no ranking of students within schools and of schools within the district. The Center formulated an overarching Comprehensive Assessment Plan (Red Pebble District T.R. Center, 2002). The Plan attempted to assess students on 12 aspects:

(1) Moral character and citizenship.
(2) Interest in learning.
(3) Academic proficiency.
(4) Intellectual development.
(5) Creativity and innovation.
(6) Problem-solving skills.
(7) Communication and collaboration.

(8) Interpersonal relationships and organizational ability.
(9) Hands-on ability.
(10) Personality and psychological soundness.
(11) Physical education and health.
(12) Aesthetics and expression.

Each student should have an Assessment Handbook of Comprehensive Qualities. Table 4.5 displays excerpts drawn from the Assessment Plan that demonstrate the district's ambition to overhaul the assessment scheme.

Echoing the district' efforts, V.P. Yu asked teachers not to test students as before. She intended to assess students using a whole gamut of perspectives that not only took into account his or her academic achievement, but also valued his or her behavior, and special talents other than the academic. Teachers and other students, thus, would have a more balanced and richer perspective of each student. She explained this as follows in the 2002–2003 School Report:

Table 4.5 The Red Pebble District assessment plan

Student self-assessment	(1) The student self-assessment is conducted once a semester. Reflect on one's strengths and weaknesses in the 12 aspects, evaluate whether progress is being made or not, and complete the self-evaluation in the Assessment Handbook. (2) Keep a record. The student should document and update their own highlights and submit the record to the classroom director at the end of each semester.
Assessment by teachers	(1) Assessment conducted mainly by the classroom director, along with teachers of other subject areas, once per semester. Assess the student's comprehensive qualities, and strengths and weaknesses, set out expectations, and give comments. (2) Assessment conducted by subject area teachers. For example, mathematics teachers should: (i) Assess the level of the student's attitudes to learning (A: Proactive and conscientious, B: Relatively conscientious, C: Ordinary, D: Unconscientious) based on homework, in-class performance, and extracurricular learning. (ii) Assess the level of academic achievement based on the final test, quizzes, practice-based homework, and in-class performance (A: Excellent, B: Good, C: Satisfactory, D: Unsatisfactory). Results should be kept in the Assessment Handbook.
Peer assessment	Students assess each other on the 12 aspects each school year, and results should be recorded in the Assessment Handbook.
Assessment by parents	Parents should read thoroughly the Assessment Handbook at the end of each semester and give feedback on their children and the classroom director. Their comments should be documented in the Handbook.

Note: The table was adapted and translated by the author from Red Pebble District T.R. Center (2002).

> We should adopt a full-facet strategy to assess students. Not only should we care about students' academic achievement, but also discover and develop students' multiple potentials, learn students' needs in growth, and help students understand themselves and establish confidence. The areas of assessment include moral character, learning ability, communication and collaboration, individuality and affection. The methods include teacher assessment, parent assessment, peer assessment, and self-assessment. There are four levels of comments: *excellent*, *good*, *satisfactory,* and *unsatisfactory*.
>
> (Merits School, 2003, p.3)

These measures from the District T.R. Center's Assessment Plan were by and large adapted, although they were not fully implemented. What V.P. Yu did change was she abolished the previous 100-point scoring system and stopped ranking students. As Teacher Hong recollected:

> We were not allowed to award specific grades to students' homework or test papers like before, this was 60 points, and that was 90, 95, or 100 points. If the work was really lousy, we were supposed to put down the word 'poor,' or otherwise we wrote down 'good,' or 'excellent.' That was said to obscure student contrast and competition.
>
> (Teacher Hong, Interview #1)

V.P. Yu named this system as *quality level + comments + special talents*, which was meant to replace the previous singular criterion, that is, test grades, to assess student learning.

V.P. Yu was sincere in the implementation of the reform. Yet, she soon encountered resistance from other administrators in the school. Such resistance led her to restrain her efforts to implement the reform.

Reform in disagreement: The restrained stage

The concerned director

The reform that V.P. Yu pushed forward sincerely seemed to increasingly disturb T.D. Wang. The discontent of T.D. Wang with V.P. Yu's action could be felt by teachers. T.D. Wang was less willing to recount what happened between her and V.P. Yu, but she recognized that she was concerned about V.P. Yu's conformity with the district's directives. T.D. Wang objected to V.P. Yu's decisions on two grounds: The district might not continue with the reform, and parents only cared about test grades.

The reform might be aborted. As the school key administrator responsible for Merits School's reputation in educational quality, T.D. Wang maintained that the restrictions V.P. Yu imposed on the hours of teaching mathematics, and the amount and form of student work would not do good to the school;

instead, they might lead to detrimental consequences. "Would it not be insane to pretend that test grades are not emphasized anymore?" she said (T.D. Wang, Conversation in the T.G. Office). One reason why she challenged V.P. Yu's decision was that she distrusted the reformers: She was suspicious of the sustainability of the reform policies. T.D. Wang asserted:

> You never know what they [the higher authorities] are thinking about... They are good at changing three times in a day (*ywe tian san bian*; *yi*=one, *tian*=day, *san*=three, *bian*=change)...In 1997, they wanted qualities-oriented education; in 1998, they wanted innovation education; in 1999, creativity education came; and in 2002, they had this.
>
> (T.D. Wang, Conversation in the T.G. Office)

She conveyed her concerns to V.P. Yu: What if the District T.R. Center administered the Uniform Examination again? What if they ranked schools again? If the district would once again use test grades as the sole yardstick to gauge a school's teaching quality, could Merits School still outperform others? T.D. Wang doubted it. She reasoned that without enough lesson hours, teachers could not even finish the required contents of the curriculum; and without sufficient homework and drill, students could not grasp the contents. Without solving a wide range of problems, students would not be able to achieve fluency and accuracy in taking tests. The school should be prepared for potential examinations. Otherwise, if test grades and school ranking were demanded again, "the school would lose its edge and reputation" (T.D. Wang, Conversation in the T.G. Office).

Parents only cared about test grades. It was the potential consequence that T.D. Wang dreaded. If Merits School lost its leading position in the Uniform Examination, she was certain that they would have a hard time enrolling new students, and more importantly, "quality students," since those "quality students" would leave for other schools that performed well in tests. The deterioration of the quality of "student sources" would activate a vicious circle – poor student quality, leading to poor test performance, leading to worse student quality, and thus worse test performance. When the school could not attract enough students, it would not be too far away from being shut down or merged with other schools!

In T.D. Wang's opinion, the very root cause of the school's apprehension of missing "quality students" was grade-minded parents. "How do they judge a school's quality? Looking at nothing but test grades!" she claimed (T.D. Wang, Conversation in the T.G. Office). Whichever school had the highest score, it was considered the best. In other words, T.D. Wang believed that Merits School's emphasis on testing was squarely because parents focused their attention exclusively on test grades. But, as the mother of a middle schooler, she understood that parents had to pay predominant heed to test grades. A high test score in schools was a powerful private capital to compete with other capitals, social, cultural, or monetary. For parents lacking powerful connections, taking tests was the last reliable resort to have their children move up the social ladder.

"This is China's social reality," T.D. Wang claimed, "You need high grades to go to a good middle school, a good high school, and a good university… competition starts from at kindergarten" (T.D. Wang, Conversation in the T.G. Office). Not different from other parents in her school, T.D. Wang expected her daughter to be admitted to the best local high school, the First City High School (a province-level key school), preferably without extra charges in addition to the tuition. There were limited sponsorship-free spots, about 300 out of 1,000 enrollees. "If her entrance examination score is among the first 100 in the district, we will earn one spot free of sponsor fees [in that high school]" (T.D. Wang, Conversation in the T.G. Office). The other 700 or so students had to pay a sponsor fee, to the amount of 8,000 to 15,000 Chinese yuan depending on their test grades. Students whose grades were lower than the cut-off point could not be enrolled, even if their parents were willing to pay the sponsorship. They had to mobilize connections (*guan xi*) to crack open the school door. Yet T.D. Wang pointed out that to be in a good school was not sufficient: Being selected into a good class was more life-determining. Based on their entrance grades, students of the same grade were stratified into different tracks: The ordinary classes (the "slow track") and the key classes (the "expedited track"). Being placed in a class packed with inferior students (*zha zi*) predicted a slim likelihood of the student's college prospects.

"[Parents like us have] no connections (*guan xi*), no money, no social status. [If our children do] not go to college, what is the way out? No job, no rice bowl. How realistic it is!" (T.D. Wang, Conversation in the T.G. Office).

Another associate director seconded T.D. Wang's point:

> To enter a university is not at all okay. It must be a top one! We can do nothing to help our child secure a decent job. My nephew wanted to be a civil servant in the province, so he took the province test. Last year, he ranked second place out of about 500 test-takers. There were three spots available. But he was out because he did not seek connections. This year, he was the third out of 1,000 competitors. His parents got wise this time. At the last step, background screening, he gave up.
>
> (A.T.D. Mei, Conversation in the T.G. Office)

T.D. Wang argued that anxious parents made it out of the question for Merits School to run counter to the social reality. The school had to passively cater to the needs of parents. She was sure that as long as the social reality, the sociocultural logic, continued in effect, Merits School would be less likely emphasize teaching over testing.

Little could be recovered regarding the reaction of Principal Li, the then principal, to the controversies between his subordinates. Nonetheless, his side was self-evident: "Simply speaking, test grades" (Teacher Wu, Group Interview #2). Perhaps, as pragmatic as he always was, Principal Li would not like to jeopardize his career either. After having been the Principal at Merits School for 3 years, Principal Li became increasingly tougher on teachers and started treating "workers" at best like "an elementary pupil" (Teacher Guo), and at worst like "a prisoner" (Teacher Chen). Principal Li demanded them to work harder. After school he would stand at the front of his office window surveying which teacher left first. Teachers who made any mistakes would be scolded in the weekly schoolwide learning meeting. It was six depressing years "in fear and trembling" (Teacher Guo) for teachers in the school. Though, in part what Principal Li did was justified in the name of assuring the quality of teaching. In 2004, he received the Renowned Principal of the City Award for Merits School's excellence in teaching and reform.

Two-pronged approach

The conflict concluded with V.P. Yu's compromise. The school administrators regained their consensus. The following semester (in spring 2003) saw V.P. Yu soften her stance and relax regulations on the curriculum schedule and reformed teaching norms. From 2003 onwards, the school was prepared a two-pronged approach. On the one hand, the school was involved in reform activities, conducted professional development for teachers, and undertook reform-focused lesson demonstrations. In its 2004 Self-evaluation Report to the D.E.B., Merits School depicted itself as a hardcore reformer. The Report summarized:

> We actively transformed our attitudes, renovated the thinking, uncovered our potentials, devoted ourselves to the reform, and exerted efforts to become a firm proponent of this grand educational reform, in which our school once again radiates the rays of humanity.
>
> (Merits School, 2004, p.1)

Seven major achievements were mentioned in the Report: (1) National and provincial experts in the new curriculum visited the school three times, teachers gave over ten demonstration lessons to visitors, and the school made two speeches at the city-level seminars on the N.M.C. reform; (2) the school collected a number of instructional cases; (3) over 100 cases or teaching research papers received national, provincial, and city-level awards; (4) the school actively participated in the district-wide exchange of reform pedagogies; (5) two school-based curricula were developed; (6) five national, provincial, or city-level research projects were conducted, for example, the state key research project, Transforming Mathematics Education; and (7) the school accomplished various reform tasks assigned by the City.

On the other hand, in everyday teaching, the school once again stressed test preparation. The enthusiasm of mathematics teachers' for reform pedagogy in regular instruction withered. They reduced the use of small groups, there were fewer outdoor activities, and student work became homogenous.

Teachers started taking a portion of lesson time allocated to physical education and arts to teach mathematics and to attend to drills.

From 2003 to 2006, as Teacher Hong recalled, teachers did not press students as hard as before, since testing and ranking were relatively not that emphasized. In July 2003, the District T.R. Center administered the paper-pencil-based final test. The level of difficulty of the problems was intentionally reduced significantly. The majority of students in the district scored 100 points. In the subsequent year, the exam paper consisted of two sections: The compulsory part and the optional part, worth 100 points and 20 points, respectively. Students could earn full marks (100 points) if they could answer all optional problems correctly. The 2005 Examination shifted gear for a third time. The Center thought that the grades could not distinguish ordinary students from students with special talents. Thus, the test was made up of the compulsory part, worth 130 points, and the elective part, worth 20 points. The grades for each part were registered separately. However, the reinstatement of the district Uniform Examination in the summer of 2006 resulted in the complete dissolution of the reform at Merits School.

The reform in dissolution: The "two-faced stage"

The Uniform Examination reinstated

The D.E.B. did not disappoint T.D. Wang. In the 2005–2006 school year, 3 years into the reform, the Uniform Examination was reinstated in the district. The foresight of T.D. Wang was paid in full credit.

In early 2006, the C.E.B. reemphasized that schools should pay attention to students' academic achievement (C.E.B., 2005). The D.E.B. said that they lacked a scientific and rigorous yardstick to measure students' academic performance, and thus announced a plan to test students selectively. The practice of ranking schools by test grades in the Examination was revived, though done in a slightly less overt fashion, because the exact ranking was not publicized via the official channel as before. Only the top five schools were mentioned during the regular Textbook Training Workshop at the beginning of the subsequent school year, but "every school knows their own standing," since "all seek that information from the Office" (T.D. Wang).

Merits School had already made full preparations for the Examination to return. Unsurprisingly, it earned the highest or second highest grades in the district across all six grade levels, which reaffirmed its reputation for academic achievement in tests. V.P. Yu was apparently pleased with Merits School's precaution too. She stressed with a note of luck and pride that some schools in the district performed awfully when the Examination was suddenly administered. Merits School truly stood out, because "we [Merits School] never loosened our grip on teaching quality" (V.P. Yu, Phone Interview). In other words, some schools went too far down the reform road and overlooked the importance of teaching to the test, but not Merits School – Merits School was rewarded for its two-pronged preparation.

From 2006, the core theme of mathematics education in Merits School returned to pure knowledge, excessive drill, and preparation for the district Uniform Examination. Ever since then, the school has normally had 13–15 lessons of mathematics weekly, in contrast to the officially mandated 4–5 lessons per week in the curriculum schedule.

The "two-faced strategy"

In November, the then principal, Principal Li, was promoted to a new position in the D.E.B. Principal Yong, a former V.P. at one local middle school, assumed the principalship at Merits School. Astutely, Principal Yong exposed the strategy that Merits School had utilized, one of course that he would continue, to cope with the curriculum reform. He stated frankly, "You gotta have two faces, one face to the outside, and the other face to the inside" (Principal Yong, Conversation in the Playground). One face was the reform face, ceremonial, pretending to adhere to the reform ideas, and intended to impress outsiders; and the other face was the real face, pragmatic, relying on heavy drill, and reflecting the day-to-day reality of mathematics education at Merits School.

To construct the reform face, Merits School appeared to conform to the provincial curriculum schedule. Take Teacher Wu's Curriculum Schedule in Table 4.1 as an example again. Officially, a full range of subject areas were offered according to the schedule. The total number of mathematics lessons per week was reduced to four. In reality, 15 lessons were used to teach mathematics, which was almost four times the official number. When it approached the last five or six weeks of the semester, the school would suspend all other subjects solely for mathematics and Chinese. Of course, on the occasions when external inspectors or visitors were coming, teachers would conform to the official schedule temporarily.

In addition, Merits School prepared a gamut of reform-aligned reports, research papers, instructional cases, and other artifacts to display to visitors frequenting the school. Principal Yong's words best captured this strategy:

> Have you been to our exhibition room on the fifth floor? Our exhibition materials are perfectly crafted. Every year, we need to greet countless inspectors and visitors. If they come to visit us, these materials will be absolutely impressive. Whatever they desire to see, it is there! Any teacher or T.R. group can easily perform a cycle of lesson study or teach one reformed lesson. Anyway, that is what you are looking for, isn't it? As for other times, we just do whatever we have to do.
>
> (Principal Yong, Conversation in the Playground)

What they had to do was teach for the test using excessive drilling.

Institutionally, Merits School had revamped and reinforced teaching norms aligned with the reform ideas at the beginning of the reform to promote the N.M.C. Yet, owing to the school's pursuit of top test grades in the district Uniform Examinations in recent years, few of the reformed teaching norms were actualized. Implementation of these norms was more to deal with inspections by higher authorities. The two-faced nature of the reform at Merits School will be detailed in the next chapter.

The two-faced strategy served Merits School well. In the past years, Merits School has scored a dozen honors. One recent award was the Exemplary School in the N.M.C. Reform, granted by the Red Pebble District, and another one was the Research and Practice Base of New Curricular Theories Award, by the State Planning and Steering Committee of Educational Sciences. Both represented the higher authorities' recognition of Merits School's efforts to reform.

Having made no attempt to challenge or topple the school's approaches to the reform, Principal Yong, easygoing and with the same pragmatism, took on the legacy of his predecessor. He reasoned on grounds similar to T.D. Wang's:

> Can you not make any difference in your school? [we ask him if he can de-emphasize tests.] [Then he stares at me as if amused by my pretended naivety.] Can you shake a tree by waving its branches rather than the trunk? No way! *Gao Kao* [college entrance exams] is still out there. It is the trunk [of the whole educational system]. We elementary schools look up to the baton of *Gao Kao*...The Bureau will not treat me nicely, and parents will devour me, if we blow the Uniform Examination this year...If we fail, the enrollment of fall will look ugly. We definitely need good students to keep our academic momentum. The quality of student sources is the key. It is not nice to say, but we can teach a monkey to climb a tree, why not a pig?
>
> (Principal Yong, Conversation in the Principal's Office, July 17, 2009)

The tree, then, was the tests, to which only "good" students could possibly manage to climb to the top. As for mediocre and "bad" students, they were the disfavored ones: They seemed to be doomed to fail in the tree-climbing contest between monkeys and pigs.

Summary

The reform was imposed on the school from the higher authorities, when not every school administrator bought into the decision. At the bottom of the school power hierarchy were ordinary teachers who had little input in decision-making and passively followed the school leaders' strategies. Throughout the reform, parents were left out of decision-making. Nevertheless, school leaders attributed the school's failure in the reform to parents' singular interest in test grades.

The disagreement between the reform-minded V.P. and other administrators led the school to both participate in the reform and prepare students for the test. Since the district reinstated its testing policies, the school ended up with a two-faced strategy to cope with the reform.

Notes

1 Better known as the Teaching Research Office (*jiao yan shi*). In January 2002, the Red Pebble D.E.B. merged the Continuing Education School for In-service Teachers with the original Teaching Research Office and regrouped it into the present Center. Nominally, the office is a nongovernmental professional agency that promotes teaching excellence and organizes professional development activities. In reality, the Office serves an extended arm of the Bureau to oversee schools and to evaluate administrators and teachers. The key members of the Office are teaching research fellows (T.R. fellow; *jiao yan yuan*; *yuan*=person), generally selected out of high-performing teachers in the local area as the teacher trainer. Respectively, there are Province, City, and District T.R. Offices. In this work, we use the District T.R. Center to refer to the agency.

2 Ms. Xinlan Ma was an elementary mathematics teacher in Beijing.

3 During the 1980s and 1990s, the length of one lesson period was 45 minutes instead of 40 minutes. There was a 10-minute break between two lessons. Between the second lesson and the third one in the morning, there was a 30-minute break for the school-wide eye protection exercise and outdoor eurhythmics.

4 Apart from the reform-minded general principles, in considerable part, teachers' desired professional conduct in instruction, from when to arrive in the classroom, to the desired language style, to how to end a lesson, were prescribed as follows:

(1) Teachers must follow strictly the curriculum schedule and the lesson plan. They are not permitted to teach without or deviating from the lesson plan and are not allowed to swap classes with each other without the permission of the T.G. Office.
(2) Teachers must wait outside the classroom one minute before class, and enter the classroom on time when the bell rings; teachers cannot be late, leave the classroom in the middle of the lesson, or run over time.
(3) As teachers stand at the podium, declare "class begins." Students should rise up, stand straight, greet each other, and then sit down.
(4) Teachers' language must be formal, concise, precise, vivid, lovely, and inspiring; the speed and tone should be at the moderate level.
(5) Handwriting on the blackboard should be neat, and the layout be well thought out; teachers should use color chalks for special effect.
(6) Teachers should attend to and adjust students' postures during reading, writing, and sitting.
(7) Teachers should not sacrifice students' learning time to deal with other students' discipline issues.
(8) In order to enhance the effectiveness of teaching, teachers should increase in-class exercises, and design varied and representative problems; homework should be assigned clearly.
(9) Teachers should declare "class ends" and await students to stand appropriately, then permit them to move freely (Merits School, n.d., p.39).

5 The format was: (1) Fold the page in half, use the left half first and then the right, and writing must be done in pencil; (2) correct all errors in the previous assignment; (3) note the date, the page number of the textbook, and the problem number; (4) write one number on one line, with one space in between; (5) leave one line between problems to write; and (6) for word problems, students need to copy the problems down in their notebook, and start a new line to work; for applied problems, students only need to write down the problem numbers, and directly give solutions; for problems with subquestions, list the questions down in order and then solve. (Merits School, n.d., p.43)

5 Teaching evaluation at Merits School in response to the reform

Dressed-up teaching norms

Lesson planning

According to Merits School's regulations, to ensure that teachers plan lessons in alignment with the new standards, the T.G. Office should conduct regular evaluation, complemented by random examination, of all teachers' lesson plans. On the surface, the school did carry out such evaluations. A system with nine reform-minded criteria was even stipulated to gauge the quality of lesson plans. The nine criteria were three-dimensional teaching objectives, emphases and difficulties, instructional design, teacher-student interaction, application of modern instructional technologies, connection with life, communication and reflection, originality, and blackboard layout.

During the spring semester, the school held two rounds of lesson plan evaluations, one on March 12 and the other on April 8. Table 5.1 shows the evaluation results of four of the selected teachers' lesson plans in the March 12 evaluation.

Following the evaluation, the T.G. Office summarized:

> Via this round of evaluation, we found that the majority of teachers prepared lessons with care, and wrote the plans in conformity with the school's requirements on lesson planning. The contents were thorough, and handwriting was neat. Most teachers showed a precise grasp of the key points of the subject matter. Instructional design was practical, and emphasized interactions between teachers and students. Lessons were extended beyond the classroom. Post-instruction reflection, communication, and re-planning were present in most lesson plans. Most teachers prepared lessons one week ahead.
>
> Several shortcomings were (1) some teachers could not precisely grasp instructional objectives, which were too general; (2) student activities were not highlighted, and some teachers only had questions without expected answers; (3) a few teachers did not individualize student work and lacked

Table 5.1 Evaluation results of selected teachers' lesson plans at Merits School

Name	*Elements of evaluation*									*Total points*	*Note*
	T.O.	*E&D*	*I.D.*	*T.S.I.*	*I.T.*	*C.W.L.*	*C.R.*	*O*	*B.L.*		
Zhang	1	1	1.8	0.9	1.0	0.8	0.9	0.7	0.8	8.9	
Fu	1	1	1.4	0.7	0.8	0.9	0.7	0.7	0.8	8.4	a
Hong	1	1	1.8	0.9	0.9	0.9	1	0.9	0.8	9.3	
Su	1	0.7	1.6	0.8	0.8	0.7	0.8	0.9	0.8	8.3	b

Note: Zhang=Grade 1, Fu=Grade 5, Hong=Grade 5, Su=Grade 5.
T.O.=teaching objectives, E&D=emphases and difficulties, I.D.=instructional design, T.S.I.=teacher-student interaction, I.T.=application of modern instructional technologies, C.W.L.=connection with life, C.R.=communication and reflection, O=originality, B.L.=blackboard layout.

[a] Overly simple, lacking reference to the Standards, and missing the Unit 1 plan.
[b] Lacking the quality analysis for the final exam of the previous semester.

> post-instruction re-planning; (4) several teachers never wrote reflections; (5) the design of blackboard layout lacked originality; (6) the application of instructional technologies was lacking; and (7) one or two teachers did not analyze the former final test.
>
> (Merits School, 2009, p.1)

To outsiders, it seemed that evaluation of teachers' standards-based lesson plans was conducted seriously, and the feedback was pertinent. In fact, those lesson plans were faked in the first place. Recall A.T.D. Teacher Zhang's case reported in Chapter 3. Her lesson plans were neatly written. In the right margin of her lesson plan notebook were written in red post-lesson reflections, though she had not taught the lesson yet. She copied those plans from a commercial publication, as others did. As observed, few teachers, if any, put their hearts into designing lesson plans, and most chose to copy from commercial lesson plan books. As various teachers stated, to them, writing comprehensive reform-minded lesson plans was more of a mechanical labor, and the main purpose was to cope with inspections from higher authorities.

Instructing

Even though teachers at Merits School learned how to teach in ways aligned with the reform, as described in the previous chapter, they did not honor the reform pedagogy in everyday instruction. Everyday instruction meant day-to-day teaching other than ceremonial occasions such as giving demonstration lessons or attending instructional competitions. The following case shows the ceremonial nature of Merits School employing the reform pedagogy and its importance.

Teacher Su was a veteran mathematics teacher at Merits School. In 2004, he was awarded the Outstanding Teacher in Implementation of the N.M.C.

Reform by the District T.R. Center. He always taught in the upper band, rotating from Grade 4 to Grade 6. In November, the head fellow of the City T.R. Office, Mr. Xu, came to Merits School for a teaching research activity. Teacher Su was one of the teachers who was asked to give model lessons. Unlike his colleagues, Teacher Su's instruction was antithetical to the reform pedagogy. As he said:

> Without those decorations and tricks like small groups or PowerPoint, we taught the way we normally teach, the most direct and efficient way. As long as students can quickly grasp the content and are able to do problems, we think it is good teaching. It is not that we cannot use the tricks well, but that we always think it is meaningless to put on a show. In the post-lesson conference, the fellow criticized that my lesson was exactly the example against the reform spirits. Later on, V.P. Yu came to apologize to me and said that was her fault, since she did not discover clearly beforehand what they wanted.
>
> (Teacher Su, Interview #2)

Teacher Su's instruction failed to please the city visitor who was clearly reform-minded. From Teacher Su's perspective, he considered the effect of reform-minded instructional approaches no more than decorative, if not counterproductive. In the pursuit of instructional efficiency and effectiveness, the constructivist inquiry-oriented pedagogy did not satisfy teachers. He could have "put on a show" like others. Together with her teachers, V.P. Yu contrived to keep Merits School's two-faced strategy in a delicate balance.

Teacher Su was not alone in thinking along that line. Teacher Fu, his colleague in the same office, Teacher Chen, and many others, all expressed on different occasions their sense of amusement, and pointed to the nullification of the N.M.C. reform. When asked how she responded to the curriculum reform, Teacher Fu said, "sticking to the one way to cope with tens of thousands of changes" (*yi bu bian ying wan bian*; *yi*=by, *bu*=not, *bian*=change, *ying*=cope with, *wan*=ten thousands, *bian*=change) (First contact). *The* [emphasis added] way was: Direct instruction without tricks like small groups or hands-on activities, coupled with an enormous amount of drilling.

Even though some hands-on activities or practices were designed at the end of each unit of the text, teachers seldom made use of them. Teacher Chen disclosed, "theoretically, they [practices] should be the focus of instruction, according to the M.O.E.'s call for qualities-based education. But, since test grades are re-stressed, and we are judged by grades, we have to abandon these activities for sure" (Group Interview of Grade 4 Teachers #2). Still, "lecturing is the dominant instructional method," and even when hands-on elements are present, they are "superficial" (Teacher Xue, Group Interview of Grade 4 Teachers #2). Since hands-on learning demanded much higher devotion of class time, and thus was less efficient, teachers were rarely willing to bother incorporating it into teaching: They had to save time for student work.

Student work

To outsiders, Merits School seemed to be serious about reforming student work. For instance, on April 23, the school T.G. Office carried out one schoolwide evaluation of student work. Each class selected two–five samples. Take Grade 1 Class 6 as an example. The work of 18 students was chosen, which consisted of the textbook, homework notebooks, test papers, and daily arithmetic practice workbooks. Table 5.2 displays the evaluation results.

Teacher Zhang, A.T.D. in charge of Grade 1 mathematics, summarized in her evaluation report:

> Several merits are (1) the frequency of work reached one time per day and the amount was appropriate; (2) students' handwriting and format were standard, clear, which demonstrated students' good learning habits; (3) the teacher marked the work in a timely fashion, students themselves corrected errors duly, and teachers re-marked the errors; (4) homework effectively reflected the difficult and key points taught of the day.
>
> Some issues that we should attend to and improve were exposed through this round of student work evaluation. First, in terms of the design of problems, too much emphasis was placed on written forms of work, while hands-on operation and extracurricular practices were scarce. Second, problems lacked innovation and originality, and were not designed by the teacher him or herself. Third, the teacher rarely gave encouraging comments.
>
> (Merits School, 2009, p.1)

No points were given apart from general comments. These comments were readdressed to teachers by V.P. Yu during the schoolwide meeting in September. She said:

> Homework and exercises for every lesson must be carefully designed, and should avoid repetitive exercises. We call for diversified homework,

Table 5.2 Inspection of student work (from Grade 1 Class 6) at Merits School

Student name	*Work source*	*Amount assigned*	*State of completion*	*State of teacher marking*	*Teacher comments*
A	Textbook problems	Moderate	Satisfying	Timely	Thoughtful
B	Homework notebook	Moderate	Satisfying	Timely	Provided
C	test Papers	Moderate	Satisfying	Timely	Thoughtful
D	Hands-on operation	Moderate	Satisfying	Timely	None
E	Daily arithmetic	Moderate	Satisfying	Timely	Thoughtful

> including basic, experimental, and hands-on problems. Having students explore and report afterwards is good. Teachers should grade homework conscientiously and in a timely manner. Teachers should put down encouraging comments. We don't suggest assigning homework to Grades 1 and 2, and homework should be no more than 30 minutes for Grades 3 and 4, and no more than 40 minutes for Grades 5 and 6.
>
> (Field Notes)

In the end, she suggested that teachers be more flexible and design a wider variety of student work to improve students' creativity and hands-on capabilities. Nevertheless, both Teacher Zhang's evaluation and V.P. Yu's suggestions contradicted what the school truly emphasized.

In the City, Merits School had long been famous for its heavy load of schoolwork. Even though the school verbally kept stressing to diversify student work and to alleviate the workload of students, their real action was excessive drilling. Student work came from four sources. The first source was called the Daily Oral Arithmetic Card (*kou suan ti ka*; *kou*=oral, *suan*=calculate, *ti*=question, *ka*=card). Usually, it was taken from a commercial workbook, each page having 32 or 64 arithmetic problems such as 21 + 9= ?, 30 cents + 2 dollars = ? dollars, and so on. Students before Grade 3 were required to finish at least one page of the card per day. Take Grade 1 as an example. On June 18, the number of arithmetic problems that most of Teacher Rui's students finished ranged from 24 to 64 problems (except for one student; he finished 385 problems in 11 minutes). On average, each student did 64 problems and spent 2 minutes finishing this assignment.

The textbook was the second source of student work. For example, the Grade 1 textbook had 114 pages, each page having five problems. The other two major sources are made up of commercial workbooks (*lian xi ce*; *lian xi*=exercise, *ce*=book), and test practice papers. Like all other grade levels, Grade 1 had two commercial workbooks, one with 114 pages and the other with 68 pages, and each page contained roughly ten problems.

The last major source of student work was test preparation papers. The number of test papers varied across the six grade levels. Teacher Tang, a Grade 1 teacher, said:

> We may give one or two test papers after each unit is completed. This is not for sure. If students don't learn the content well, we will do more. Generally, it is one paper in one day or two days. Every test is marked. If students make errors in one test, we print out the test paper and ask them to redo it.
>
> (Teacher Tang, observation in Grade 2 Office)

Since Grade 3 and Grade 6 were the focal grades in the district Uniform Examination, they assigned more work. For example, from late February to early July, Grade 1 used six volumes of test preparation papers, each volume consisting of 15–20 sets. In general, each set had four pages, contained about

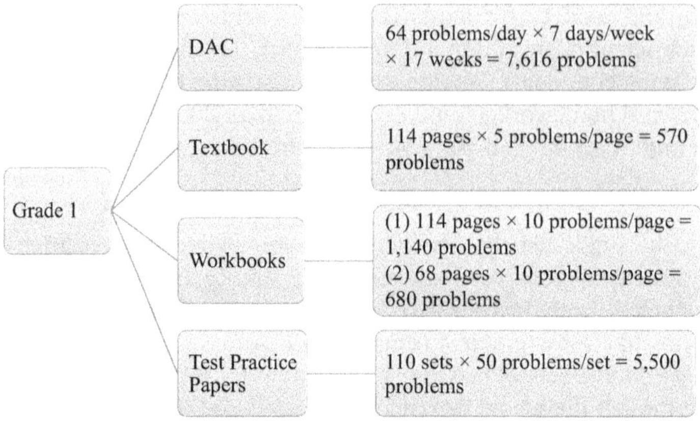

Note: DAC=Daily Arithmetic Card.

Figure 5.1 The source and volume of student work at Merits School.

50 problems, and should be finished in 90 minutes. T.D. Wang counted how many test papers her Grade 6 students had practiced by July 1, two weeks before the final test. The number was 131 sets.

Figure 5.1 charts out the sources and the average amount of student work assigned to Grade 1 students during the spring semester, over a span of 17 weeks.

According to the chart, a Grade 1 student generally had over 15,000 problems from March to July. The amount of time he or she spent on mathematics in school and after amounted to at least 150 minutes per day.

Needless to state here, the above figure is an oversimplified illustration of student work, which can hardly capture the great variety of problem types[1]. Among the problem types, some were limited to basic arithmetic operations, while some required higher-order reasoning and multiple procedures. Different types of problems entailed different skills of students and had varied time demands. But, the essence conveyed by the diagram is clear. That is, students at Merits School had a considerably heavy workload.

The ideas of diversifying and differentiating student work were aborted too. One practical difficulty was the large class size (60 students on average per class) at Merits School. It became impossible for teachers to design student work based on individual student needs. Teacher Rui commented:

> Differentiating student work is good in imagination, but definitely not a realistic idea for a large class. First of all, it is not feasible to write down problems on this piece of blackboard. For a class having more than 60 students, about ten are outstanding students, 30 good, and 20 more

> so-so. Tell me how to lay out student work? This section for the outstanding, that for the good, and that for the ordinary? The problems you design for outstanding students are tough for sure. They cannot work out every problem, right? You need to help them out, and need to coach them, right? Ask the outstanding students to circle together? Okay, you ten come here. Speak to them softly, the students in the rear cannot hear; louder, you disturb others. Besides, what about the other 50 kids when you teach these ten? What about the 20 poor students? Leave them alone? They cannot concentrate on their own problems at all if so. Besides, those left behind are in fact the main problem, the main target of us. By the way, where do we find the place for them? The whole 40 minutes are wasted on regrouping them. The class will become chaos. If we had 30 kids like the U.S., each ability group having about ten kids, then that is fine, that is possible to personalize homework and exercises.
>
> (Teacher Rui, Interview #2)

V.P. Yu's call for diversified work, including basic, experimental, and hands-on problems, meant to provide every individual student with equal and ability-appropriate opportunities. But, large class sizes, coupled with teachers' focusing on test preparation, prohibited teachers from adopting this well-intentioned reform initiative.

Student work was the most important and heaviest component of students' school life, and so it was for teachers too. The core learning and teaching labor did not occur in the 40-minute lesson, but in the excessive amount of student work. Teachers had to grade all work. As observed, Merits School's teachers generally spent 3 hours grading student work on a daily basis. As Teacher Zhang testified, "we have 64 students this year. Usually, grading one student's work takes me at least 2 minutes. You calculate, how much time it is in total?" (Teacher Zhang, Interview #1). The time she spent grading student work was almost doubled in the last six weeks of the semester. She felt as if day after day all she did was grade students' homework, this monotonous, time-consuming and energy-demanding task. Her colleagues shared similar sentiments. A very depressing morale permeated in the school.

The most direct consequence of heavy student workload was revealed in the reality that none of the teachers who were interviewed liked their work. No teachers we interviewed wanted to be a teacher if there were a second choice. Teacher Su came up with a jingle popular among Chinese teachers. He said it perfectly expressed his feeling: "Getting up earlier than a rooster, going to bed later than a hooker, earning less than a hawker, eating worse than a hog, and working more than an ox." The jingle described teachers in terms of animals and equated them with laborers.

Teachers' emotional responses to the curriculum reform, as well as the teaching profession, ranged from disappointment to tiredness, to dislike, to anger, to numbness. Teacher Zhang and Teacher Rui felt powerless regarding the resurrection of high-stakes tests. Teacher Tang outspokenly declared that she hated being a teacher. She considered her job slightly better than a pedicab driver. To maintain her class' competitive ability in tests, she admitted that she had to torture her students by drowning them in an ocean of student work, which was against her conscience, however. Teacher Hong was clearly angered and sounded in despair: "Reform! Reform! Why the more they reform, the worse it gets?! China's education is completely hopeless!" (first contact). Even though she embodied some essential elements of the reform pedagogy in her teaching, she had to assign and handle the same type and amount of student work as others did.

To the extreme, Teacher Su saw himself as "numbed." Like Teachers Fu and Wang, Teacher Su had also gone through several educational reforms that were under different banners. But, all those well intentioned reforms were aborted halfway. His disappointment at recurrent reform attempts was so evident that he was convinced, "whoever believes in them [the higher-up reformers] is doomed" (first contact).

In China, a well-known metaphor is used to describe teaching by excessive drilling and practice: The "ocean of problems tactic". The purpose of the tactic is straightforward. As one of the associate directors at Merits School said, "we cannot guess what kinds of test problems the T.R. Center is going to design, so [we have to] drill extensively and wholly" (Ms. Xiu, Interview #1, T.G. Office). In so doing, students become familiarized with the whole gamut of problems to the greatest extent, thus enhancing their potential to perform well in the final examinations.

Teachers themselves did not support this tactic, either. They sounded strongly resentful of the fact that they had to force students to do so much work. Teacher Chen was extremely sympathetic to her students: "For such young children, you ask them to do one workbook after another. Sometimes, their parents even buy extra ones for them. Anyway, we feel it is so heartless, so cruel...They are exhausted, so are we teachers" (Teacher Chen, Interview #1). But, since the school valued test grades more than anything else, she admitted, "all peers around you are doing it that way, you have to follow the majority too" (Teacher Chen, Interview #1). Otherwise, "students may become happy eventually, while the school gets upset at us" (Teacher Tang, Observation in Grade 2 Office).

In order to have students treat student work seriously, students who failed to correct the errors made in their work would face a heavy penalty: Leaving one error uncorrected might mean redoing the problem ten times. On September 27, Teacher Su retained four students after school:

> They did not correct the errors they made last night. All were punished to redo the uncorrected problems ten times – copy the questions in their notebooks and carry out the calculations carefully. Teacher Su says, "The questions are not hard to correct. They only needed to put down the right answers on the margin. But they did not. If students know how to correct the problems but do not, then we must punish them. It is appropriate

> to teach them a memorable lesson." Teacher Su stresses, "The very purpose we grade student work is to identify errors and to have students know what their weakness is. Otherwise, teachers will not labor to grade it!" One girl finishes her work and is walking out of the classroom. We ask her, "Any effect? What effect?" She giggles guiltily, "Too profound an effect! From now on, we will never dare not to correct errors!"
>
> (Field Notes)

Teachers did not like this approach either, but apparently, they used it every once in a while, most probably to deter rule breakers.

Despite the heavy workload at Merits School, some parents even purchased extra drill workbooks for their children. One Grade 1 student's father bought an additional commercial workbook for him. On June 16, he wrote a note to Teacher Zhu, the mathematics teacher, stating that:

> As time approaches the final, in order to have Huan [the name of his son] soundly grasp basic knowledge, we selected a volume of test papers to supplement the school's practice and general review. Today, he finished independently one set on money, time, and subtraction within 100, and we graded his work…To collaborate with the school, we hope to help Huan improve his academic achievement as soon as possible, and to have him obtain an outstanding grade.
>
> (Parent's Notes)

That test paper had five pages and 80 problems. Like Huan's father, many parents at Merits School managed to collaborate with teachers in similar ways. And their expectation for their children was simple, that is, "to earn a good score – 100 points –the best" (Teacher Tang, Observation in Grade 2 Office).

Some parents indeed disapproved of the ocean of problems tactic. In June, one month away from graduation, a father of a Grade 6 student at Merits School approached the Principal of Pioneer School and said he wanted to transfer his son there from Merits School. He complained that his son had to stay up late till 11:00 pm to finish homework. It was not one time, but night after night. Having no other alternatives, he attempted to seek a way out of the school. Disregarding homework was not a choice. If students failed to turn in their assignments, they were very likely to be scolded in class and parents would be called in to be cooperative. Such anecdotes were not rare at Merits School. Some parents chose to mimic their children's handwriting and did the homework instead in order to have their children rest earlier.

In this regard, parents at Merits School had little room to challenge teachers' authority. A few years earlier, as one parent interviewee informed us, one mother attempted to ask her daughter's Chinese teacher to assign less homework. The teacher, a veteran in the school, denied her request, saying that it was the school's tradition, and she could take her child out if she was feeling unhappy. The interviewee went on to add, "We only heard from others that Merits School has excellent teaching quality, but never knew what resulted in the quality before really having our children sent here" (Mother 1, Interview #1). On most occasions, parents were hesitant to give critical feedback to teachers, never mind to defy teachers. "Your child is in the hands of teachers, who will [do that]?" Another interviewee concurred (Mother 2, Interview #1). Thus, owing to their subordinate status, parents usually attempt to maintain a seemingly amicable but, actually, utilitarian connections (*guan xi*) with teachers.

Teachers disclosed that, without the acquiescence of the school administrators, they did not dare to assign an excessive amount of student work. The comments on the inspection of student work mentioned in the previous section appeared self-critical, and the suggestions were genuinely with the reform in mind. Yet, they were written more to impress outside inspectors and visitors. The insiders were conscious of their own depressing and heartless reality. Though, in order to preserve their academic superiority in the district, Merits School was unlikely to abandon its ocean of problems tactic.

Tutoring

As observed, the T.G. Office asked teachers to turn in their tutoring plans and logs for inspection in late June. But when every student was given the same work, tutoring according to students' abilities was essentially unattainable. In general, each class had five to ten "bad" students (officially, they are called *xue kun sheng*; *xue* = learning, *kun* = difficulty, *sheng* = student) to pull up. Teachers said that they generally put their greatest efforts in those lowest 10% of students, who most detrimentally affected the average test grade. Teachers did set up records for those low performers and gave them fewer and easier problems; however, they could hardly spare time to provide other work for other students.

Tutoring low level students was increasingly stressed at Merits School, particularly from 2006 onwards. But a practical issue that teachers faced was finding the time for tutoring. Out of safety concerns, higher authorities did not allow schools to keep students for tutoring after school. Thus, teachers had to use the time between lessons to tutor students. As Teacher Hong said:

> We have to seize those poor students tightly at school. Oftentimes, their work is already fraught with mistakes when one lesson is over. So they will be asked to stay in the classroom to correct errors during breaks. Throughout the whole morning, they don't even have any time to go to the toilet. The good students are relatively better off. They make fewer errors, and may have 10 minutes free.
>
> (Teacher Hong, Interview #1)

Teacher Hong pointed out one of the unintended consequences of tutoring: Making visible the differentiation between "bad" students and "good" students. While "good" students could still enjoy a moment free, "bad" students were even deprived of the opportunities for recess due to their low academic performance. It also left one to wonder how low level students were viewed by their peers because of the different treatment.

The tutoring records could not truthfully reflect how teachers conducted tutoring. Teacher Chen commented:

> We know the state of our own classes. Indeed these students are with learning difficulty. But the real process is unlike what is written here. It says we did this and that on Monday or Wednesday. That is not for sure. Real tutoring is very flexible. For example, if a student's homework has errors and problems, just tell the student where the problems lie in and ask him to redo the work. Essentially, this is tutoring.
>
> (Teacher Chen, Interview #2)

The school's inspection of tutoring records had little substantive meaning to monitor or guide teachers' behaviors. Again, it revealed the ceremonial nature of the school's reform-aligned teaching norms.

Assessment

Without using any reformed approaches, assessment of mathematics teaching and learning at Merits School was based purely on paper-pencil tests. In particular, the school placed a strong emphasis on the district-wide Uniform Examination in summer. Teachers of the same grade level were ranked according to student average grades, and this ranking was made known to every teacher in the group. In the summer district Uniform Examination, Teacher Zhang's Class 6 (Average = 93.44) and Teacher Rui's Class 1 (Average = 94.25) ranked the sixth and fifth, respectively, as Table 5.3 shows.

After the final test, teachers should conduct the quality analysis of the final examination in the following semester. The analysis should analyze issues revealed via the test, and envision remedial strategies. Teachers should include the report in their lesson plan notebooks. One illustration of this report is given in Table 5.4.

In this report, the teacher identified three problem areas and four steps she could take to improve her teaching.

Test grades once again became the definitive yardstick of the district to assess schools. In response to the district's policy change, for Merits School, securing its ranking in the district test became the sole goal. Teacher Hong voiced her anger:

> The T.R. Center only relies only on a test paper [to assess students]! It says to strengthen the reform, but in fact it demands test grades more and more. Before, ranking was not permitted, Uniform Examinations were not allowed, and all evaluative comparisons were abolished. In 2006, everything is coming back – Uniform Examinations, ranking, and

Table 5.3 Average grades of Grade 1 in the Uniform Examination at Merits School

Class	*Average*	*60–70*		*71–80*		*81–90*		*91–100*		*Average difference*	*Rank*
		No.	*%*	*No.*	*%*	*No.*	*%*	*No.*	*%*		
1	94.25	–	–	–	–	10	15.38	55	84.62	–0.60	5
2	95.38	–	–	–	–	7	11.11	56	88.89	0.53	3
3	95.72	–	–	–	–	4	6.35	59	93.65	0.87	2
4	95.94	–	–	–	–	5	7.81	59	92.19	1.09	1
5	94.37	–	–	–	–	13	20.63	50	79.37	–0.48	4
6	93.44	–	–	–	–	11	17.19	53	82.81	–1.41	6
Grade level	94.85	–	–	–	–	50	13.09	332	86.91	–	–

Note: Full marks=100 points, 60–70 refers to the number of students whose test grades fell between 60 points and 70 points, average difference=class average – grade level average (94.85).

Table 5.4 Quality analysis of the final examination

Grade 1 Class1 Spring term	*No. of students*	*No. tested*	*90–100*	*80–89*	*70–79*	*60–69*	*<60*
	58	*58*	*50*	*8*	–	–	–
Problems	(1) [Students were] confused between more and less, could not distinguish "more than" from "less than". (2) Too young to read to understand the questions asked. (3) Speed too slow.						
Remedial measures	(1) To strengthen training regarding more or less. (2) To strengthen training in speaking in full sentences. (3) To reinforce classroom organization and management, and to strive to cultivate good listening habits. (4) To lecture essential knowledge points and practice more, and to strengthen lesson preparation.						

> obsession with test grades regardless of anything else. Can [the Office] not do anything useful? Simply exhaust children. Why not have students explore more, play more, and practice more! Grades, grades, students are exhausted, and teachers are stressed out.
>
> (Group Interview, Grade 5 Office)

All reform-minded assessments that curriculum reformers promoted yielded to the high-stakes testing. The district's policy shift triggered a ripple effect: "The district demands grades from the school, the school demands grades from us, and then we have to demand that from students" (Teacher Chen, Interview #1). Naturally, the ocean of problems tactic in teaching was put into effect. T.D. Wang, also a Grade 6 mathematics teacher, said that students had to do two, sometimes three, sets of test papers daily from May to July.

In the summer test, Merits School's Grade 6 averaged 91.07 compared with the district-wide average of 85.30. But the school lost first place to one of its competitors by 0.34 points. Because one student in T.D. Wang's class only got 2 points due to his "I.Q. problem," the whole school was adversely affected.

For regular teachers, then, not being overshadowed by their colleagues was critical. Otherwise, the Principal might have a conversation with teachers scoring low. In the test, Teacher Zhang was not very happy: She ranked last in her Grade 1 group. In the following fall semester, she was asked to see the Principal who requested her to catch up. The approach of ranking against peers created high pressure on teachers. One night, both Teacher Rui and Teacher Zhang worked late till 6:30 pm:

> We ask Teacher Rui how her class did in the past district test. Pointing to Teacher Zhang and herself, she says, "She was No.1 and we No.2 – from the bottom up. So we are making efforts to catch up this semester." "With whom?" "The other four teachers." We probed her further, "Is it that different to rank top?" She stops grading, "In essence, it is not. Even if there is a difference as large as 5 points between two classes, that does not prove that you are a better teacher and I am worse. There are a whole range of factors that affect student grades. But, every final test, we are ranked horizontally among six teachers: Whose class is number 1 and whose is number 2…it is made very clear to everyone, and the school and parents are highly attentive to that. The ranking exerts a heavy pressure on you. If your class is the lowest, you will feel ashamed about yourself – people will look down on you, questioning underneath why others can obtain high grades and you cannot. And you will be thought of as incompetent. I don't want to lose face. How humiliating it is." (Field Notes, Grade 2 Office)
>
> Parents give us much invisible pressure too. We have six classes at the same grade level. They are picky about which class to place their children in. Say, if we don't care about test grades and instead cultivate students' real abilities, and my ranking remains low over 6 years, no parents would like to have their kids learn with me in the next cycle. Maybe other classes are packed with 60 to 70 students; my class may only have 40 something.
> (Teacher Zhang, Interview #3)

The ranking scheme literally took away teachers' professional authority, and left them in a vulnerable and insecure state. Because the relative rankings against their colleagues were so unstable, teachers' sense of professional competence and confidence suffered constant self-doubt and challenges. To preserve or redeem their professional integrity and self-esteem, beating colleagues in test results seemed the only way out.

As teachers were once again absorbed in pursuing test grades, colleagues were turned into rivals. Teachers themselves were clearly conscious that they were watched by colleagues in the same group, and in return they had to keep a close eye on their rivals. One Thursday late afternoon, only Teacher Chen was in the office:

> You will not find anything meaningful, if you ask us all together. No one will really tell you how they teach in front of others. We are very careful about our mouths when talking in the group. It is not always that sort of heart-to-heart communication between us. All guard against each other. Probably you haven't had any working experience yet in real society, real schools, right? If you had, you would understand what we mean. We are not that harmonious like what you imagine. We always compare and compete with one another in the same group, sometimes in the school… Collaboration is all on the surface and maybe only 20% of our time is spent on such collaboration…we rarely have deep interactions. No one is willing to be the lowest in final exams. Particularly in such an educational system, schools compete with schools, students compete with students, well, Heaven blesses us if students really compete! The problem is teachers have the bitterest competition! It looks to you our six colleagues are very nice and in great harmony, right? But, deep down we are all battling. We are like in a war. So, in T.R. group learning time and discussions, you don't really show your aces to others, like how much you know about this knowledge point, how you handle this content, what exercises you are giving to your students. Of course, when higher-ups are present in these meetings, we have to show our best faces…then, you are awed by how much they know the content and you see the real gap [between you and other colleagues]…What are we competing for? What else do we care about? Test grades.
>
> (Teacher Chen, Interview #1)

The enemies were their own colleagues, and test grades were the trophy that they competed for. Collegial communications were dampened. The feeling of shame shoved teachers into rivalry situations. More precisely, it was the fear of being left behind that propelled them to enter into the grades race. Unfortunately, if a teacher was left behind, the only way to redeem her or his esteem was pressing the students: Giving more lessons and drilling more.

For students, test grades differentiated the elite students from the low performers. Students in the top rankings were preferred in and outside school, and considered good people morally and behaviorally, while those with poor test grades were referred to as "bad students" and less favored. Students themselves were conscious of who were good students and who were poor in the classroom. For example, one boy in Teacher Hong's class described, "We good students had no problem with the fast pace of this lesson, but those poor students could not make it" (Student A). Students with intellectual disabilities and those who were developmentally and cognitively slower than their peers were even more disfavored:

> Even regular students whose brains respond slightly slower can hardly make it, not to mention those with disabilities. They can only pull down the average grades. Isn't the Grade 6 student in Wang's class the case? Whoever encounters students like that is unfortunate. Honestly speaking, it is a misfortune. But, those students are also counted in when they calculate the average. If the grades are low, they think you are incompetent.
>
> (Teacher Zhu, Observation in Grade 1 Office)

Test grades even affected the life of teachers, students, and parents in more nuanced fashions because of the strong social connotation. As A.T.D. Mei, a Chinese teacher, recounted:

> The final exam in the winter is extremely annoying. Following the exam is the winter vacation. It is time for the Spring Festival. During the Festival, relatives keep asking how your child did in the test. If your child performed awfully, without doubt as parents you would feel humiliated. Then you will not have the heart to treat the child nicely. The whole vacation is probably spent in misery. It is the same for teachers. If your class ranks low, you do not even have the mood to celebrate the festival.
>
> (A.T.D. Mei, Conversation in the T.G. Office)

Due to parents' mentality of expecting their children to excel in tests and to win out in the competition for decent jobs, they suffered disappointment and even humiliation when their expectations were not met. Understandably, such expectations might be frequently conveyed to their children. Hence, obtaining high grades in tests was particularly important for children to please their parents.

For some children, it might be where their motivation came from, and the major reason for them to be committed to learning. The researcher's brief conversation with two Grade 6 students in Teacher Su's class underscored this point:

> [Why do you work on so many mathematics problems?] Boy 1: So that we can do well in mathematics tests! Boy 2: Learning more knowledge so that we can perform better in the final. [Is it that important to do well in tests?] Boy 1: Yes. Because my parents will be happy then! And I can be a more knowledgeable person when I grow up. Boy 2: Finding a job! We think only those who do well can have a good job! [What does it mean to do well in tests?] Boy 2: Above 95 points. Boy 1: Best earn 100 points. [Is scoring 100 important?] Boy 1: Very important! Then my parents will be pleased, and we can also find a good job in the future.
>
> (Field Notes)

Literally, the pressure for "doing well in tests" activated a chain of *circular torturing*: "Parents torture the school [administrators], the school [administrators] torture(s) teachers, teachers torture students, and students have nobody to torture but go back and torture their parents" (A.T.D. Mei, Conversation in the T.G. Office).

As shown above, test grades were the singular standard to judge the performance of the school, teachers, and students at Merits School. To the administrators, test grades concerned the school's competitive ability and survivability; to teachers, higher test grades suggested that they were better than their peers, thus preserving their professional dignity and saving face; to parents, high test grades were associated with their children's ability and prospects to secure success in life. Conversely, having a child perform poorly in academics was humiliating; to students, test grades defined who they were. Under this heavy test-centered culture, reform-minded assessment could hardly find its place at Merits School.

Summary

Merits School institutionalized a reform-minded system of teaching norms that were revised in accordance with the reform spirit. In practice, the school carried out formal evaluation of teaching simply to cope with inspection by higher authorities. Since the school focused on raising test grades, heavy drill became the norm in mathematics teaching and learning. The pursuit for high test grades was the most important aspect of schooling for teachers, parents, and students. The teachers acknowledged that such a pursuit was detrimental to the emotional and physical well-being of teachers and students in the school.

Note

1 It would also be wrong to assume that those students at Merits School were doing *parrot mathematics* (O'Brien, 1999) in the sense of performing numerical and symbolic manipulations by rote memorization without any understanding. On the contrary, the mathematics that those students tackled had high sophistication and required profound conceptual understanding and a high degree of proficiency. They were tough problems.

6 Teachers' learning and professional development at Merits School

The teachers' change

As described earlier, administrators and teachers at Merits School approached the N.M.C. reform with a two-faced strategy. To outsiders, the school managed to display its successful implementation of the reform. However, inside the school, teachers were purely test driven. As a consequence, reform-aligned lesson plans and records were fabricated, teachers' instruction rarely used reform pedagogies, mathematics lesson hours were almost three times more than the official schedule, and students suffered from heavy drilling.

Nevertheless, as far as their own learning and growth, teachers in the school had acquired something positive out of the reform. Teachers were exposed to new thinking, gained new ideas, and learned how to use reform pedagogies. At Merits School, all mathematics teachers, especially teachers who started their career around the initiation of this reform, voiced their recognition of the reform's central tenet, that is, teachers serving as knowledgeable facilitators, leaders, and cooperators, and students actively engaging in collaboration and inquiry. Furthermore, most teachers acknowledged that they had acquired several reform pedagogies. Even teachers as sarcastic about the reform as Teacher Su and Teacher Fu admitted that they were favorable to the fundamental ideas of the reform. As far as *instruction* in a 40-minute lesson was concerned, it was observed that teachers paid a lot of attention to students' input and ownership of learning. For instance, Teacher Zhang and Teacher Hong frequently encouraged students to express their own views and attempted to involve students in small group-based, hands-on activities. These changes represented a considerable departure from the traditional attitudes towards and practices of teaching and learning mathematics, which tended to be teacher-centered.

Changed beliefs

The development of instructional languages was the most immediate reflection of the fundamental change of teachers' role in and beliefs about learning and teaching mathematics. Different from the traditional mindset that students should listen to and obey the teacher, students and their explorative activities were foregrounded in teaching. Students' views were actively solicited. Teacher Rui summarized:

> The most common sentence we now use is: What have *you discovered*? This is most frequently asked. "What have *you found*? What have *you made sense of*? What questions can *you raise*?" are all contemporary terms. We didn't have such language before.
>
> (Teacher Rui, Interview #1, emphasis in original)

By attending to students' discovery, sensemaking, and individual perspectives, they promoted a more participatory, explorative, and equal culture in teaching and learning.

Teacher Hong's lesson that used small group discussions frequently supported the validity of Teacher Rui's observation. Take one episode for example:

> It is a lesson on reciprocals. The definition of reciprocal had been introduced in the previous lesson. Teacher Hong asks students to identify which two numbers produce 1 – 3/8, 5/4, 3/5, 7/10, 4/5, 2/3, 10/7, and 8/3. Students quickly point out the pairs of 3/8 and 8/3, 5/4 and 4/5, and 7/10 and 10/7. Then she goes on to probe, "Remember the definition of reciprocal – if the product of two numbers is 1, then the two numbers are reciprocals. Which numbers are reciprocals mutually (*hu wei dao shu*; *hu*=mutually, *wei*=being, *dao shu*=reciprocal)?" Students raise their hands. A male student is called up and responds, "3/8 and 8/3 are reciprocals." The teacher emphasizes immediately after him, "3/8 and 8/3 are reciprocals *mutually*. That means, 3/8 is the reciprocal of 8/3, 8/3 is also the reciprocal of 3/8. The language of mathematics must be precise." After students name the rest of the combinations of reciprocals, Teacher Hong proposes, "Now, discuss with your neighbors and explore what characteristics you have found regarding the reciprocal of a fraction?" Quiet in the room is broken right away. Students talk with their peers for about 30 seconds as Teacher Hong claps her hands to call them back. Gesturing towards one female student, she asks, "What have you discovered?" The student stands up and answers swiftly, "Simply speaking, the positions of the denominator and numerator of the fraction are interchanged."
>
> (Observation #2 in Grade 6)

The above case was simple, but it had some merits worth noting. Teacher Hong did not tell students which two numbers were reciprocals, but had them identify the numbers that produced one first, thus referring them back to the definition of reciprocal. In this way, Teacher Hong helped reinforce students' understanding of this mathematics concept. She also allowed students to communicate and explore with their peers the characteristics of reciprocal fractions. In the 40-minute lesson, students had small group discussions six times.

Engaging students in learning

As Teacher Rui suggested, in traditional mathematics classrooms, teachers were the unquestionable authority, and they demonstrated examples out of the textbook, while students listened passively and mimicked afterwards. "Traditional teachers tend to open a new lesson by saying at the front that today, we will have a lesson on example X, on page Y" (Teacher Rui, Interview #2). Traditional approaches gave students few opportunities to express their own views or to collaborate with peers. After the reform, Teacher Rui emphasized that teachers would present to students a well thought out and interesting scenario or story through animations or PowerPoint demonstrations, out of which students were led to explore and decipher the latent mathematical information:

> Now, when we give an example, basically teachers do not copy the textbook. We will design a scenario, an interesting scenario, and then lead students to enter into the situation. It is like telling a story, and we naturally come up to the example. For instance, we show students a picture through the projector, with lots of fruit trees, ten or so apple trees, five pear trees, and so on. Before, teachers would directly tell, "Look at the picture, there are ten apple trees and five pear trees, how many trees are there in total in the picture?" That is before. Now, we ask questions like, "What information can you find from the picture?" "What questions can you ask?" Actually, kids cannot make the point at first. Some would say, "we found the sky is very blue," "we found some children are playing," and the like. At this juncture, teachers need to step in and ask heuristically, "What mathematics problems can you find?" "What mathematics knowledge does it have?" Or, more specifically, "What addition questions can you ask?" In this way, students can instantly grasp the core of the problem. Thus, questions are come up with by students themselves instead of being given readily by teachers. Then, when they solve the problems, they feel like they are solving their own problems. So, students feel particularly engaged. It is much better than giving students a problem and having them solve it passively.
>
> (Teacher Rui, Interview #2)

Via activities such as reading a picture, students were provided with multiple learning opportunities. The traditional pedagogy fails to connect mathematics to children's real lives, and renders it impersonal and less meaningful to children. The new approach involved students in observing the scene, extracting key information, and learning to express it in mathematical terms. In doing so, students could learn to abstract a concrete life situation into mathematics. Teacher Zhang agreed with Teacher Rui:

> In traditional instruction, the first step was reviewing the content learned last lesson so as to lay the stepstone for this lesson. [The drawback is

> that] immediately, students were confronted with knowledge and their interest waned. We now emphasize to create a situation for students so that they don't feel this is a mathematics lesson. On the lower bands, we make up stories like exploring the Kingdom of Numbers, or visiting the Kingdom of Mathematics, etc. Younger kids love that type of story. For older students, we will raise a mathematically elegant question, like a brain teaser, a joke, or an intelligence quiz and the like. When teaching the topic on the change of the surface area, we designed a problem, in which we asked students to get eight pieces of cake by three cuts. Some knew the answer right away, and some really thought hard to figure it out. [To get eight pieces,] you perpendicularly slice the cake twice in an X shape and then have one horizontal cut across the cake. Students' interest was pepped up right away.
>
> (Teacher Zhang, Interview #3)

Learning by inquiry

One of the reform strands is learning by inquiry. It is recommended that teachers take advantage of students' prior knowledge to help generate new knowledge, and that teachers avoid telling students the knowledge or answers directly without having them observe and reason in the first place. This was one of the changes that several teachers mentioned. Teacher Rui talked about this at length:

> Student inquiry is greatly emphasized nowadays. Basically, we do not give any knowledge readily. Before, teachers would directly tell students what the formula is, for example, the formula of the area of a triangle is the base times the height divided by two. No one bothered knowing how it was obtained. Now, we must have students inquire by themselves. When we teach the lesson, we ask students to revisit what shapes we have learned before, like parallelograms, squares, and rectangles. We have learned how to calculate the area of a parallelogram, square, or rectangle. In fact, the area formulas of the latter two types of figures are both the length times the width, and that of the former is the base times the height. Then we ask students to think that today we are experiencing a new figure, triangle, and we want to know its area. We may ask them, "With which figures do you think triangle appears to have a relationship?" We have students prepare two same triangles or set squares in advance. Now ask them to regroup the triangles, and report on their findings. It is very clear that the composite figure will be a parallelogram. That is, a parallelogram can be divided into two triangles. Since we have learned the area formula of parallelogram, then the following will be straightforward. All the above words must come out of students' mouths. We cannot tell them directly. If it is said by us, it is exactly "spoon-feeding." We must have students themselves come up with these findings and conclusions.
>
> (Teacher Rui, Interview #2)

Using hands-on approaches and cooperative learning

Another major pedagogical change was employing hands-on and cooperative learning. Particularly in the first 3 years of the reform, students had had many hands-on and small group-based activities. Teacher Zhang had this to say:

> Teachers do not tell them to memorize this or that. If so, students will lose interest completely…When we taught measurement, we asked students to prepare all the rulers they could find, all kinds of measuring tools, like ropes, meter tapes, diameter tapes, meter sticks. We led students to the playground and let them measure items they could find. They did that first independently. If it could not be done by one person, small groups would be formed. Afterwards, we came back to the classroom and had students report on their findings. You would find their gains were very, very fruitful. Some students would say if we measure long things, for example, the racetrack, we need to use a tape. A ruler would be too short; if we measure a book, then we need to use the ruler; if measuring the height of a student, we would use a meter tape. They come up with all of these.
>
> (Teacher Zhang, Interview #3)

The emphasis on hands-on practices and application had led to a one-day school-wide cross-discipline activity, the Trade Fair of Merits School. It was held on November 23, 2004 as a comprehensive event for students to practice social skills and apply knowledge of mathematics. At the fair, students from different grade levels were asked to bring some objects for sale, while other students attending the fair learned to bargain, calculate, measure, use money, and other skills. But, because of the pressure of the Uniform Examination, as Teacher Hong informed, it was now hard for teachers to find time to take students outside classrooms for such activities. Even so, some reform spirit was preserved in their classroom instruction. We paid a visit to Teacher Feng's Grade 2 classroom. Like Teacher Hong and many others, Teacher Feng asked her students to bring objects found at home into the class when she gave a lesson on millimeters. Some of the objects were paper clips, bank cards, toothpicks, coins, and the like. In a 38-minute lesson, she organized 11 hands-on activities that consisted of guessing the length of one meter and one centimeter; finding one centimeter on the ruler; measuring the length of the mathematics textbook individually, by a group of four students and a group of two; and feeling the length of one millimeter. The total time that students spent on hands-on activities amounted to 13 minutes, or one third of the lesson time. However, one boy sitting beside us disclosed that they did not have such activities regularly. Perhaps, our presence prompted Teacher Feng to teach the reform-minded lesson. Yet, the episode still demonstrated that Teacher Feng was capable of using hands-on approaches to teaching.

Hands-on approaches to learning mathematics proved to be an essential shift from the traditional image of mathematics learning. In Teacher Zhang's words:

> Recalling when we were in elementary school, never did we have teachers ask us to bring so many manipulatives. Whatever the teacher lectured on to us, we should memorize it. If they told us 1 meter is equal to 100 centimeters, then that is what it is. Never were we allowed to have hands-on activities.
>
> (Teacher Zhang, Interview #3)

She could see in person the advantages of using manipulatives to teach and learn mathematics. At the most superficial level, children were apparently more engaged and their interest and attention improved. More fundamentally, hands-on activities "translate the abstract into the tangible, the felt, the touched" (Teacher Zhang, Interview #3) so that children could construct the meanings on their own. Thus, children's knowledge of mathematics was not detached from their lives, but about their lives; for example, "the number of 1 was connected to one apple, one pencil, or one child" (Teacher Zhang, Interview #3) instead of being meaningless to them.

The above description shows that substantive learning had indeed occurred to teachers at Merits School. Such beliefs about mathematics teaching and learning marked those teachers' shift from the traditional pedagogical positions: Teachers as dominant directors and students as obedient followers. Indeed, some of them seemed to have tried very hard to materialize their beliefs in their classrooms. Yet, we have to bear in mind that their application of reform pedagogy was so limited that they could hardly compensate for the detriment which other key teaching activities, that is, using excessive student work for test preparation (that is, the ocean of problems tactic), caused to students and teachers in the school.

Professional development for teachers

Teachers' learning and development did not happen overnight. To promote reform pedagogies, teachers at Merits School were involved in a mandatory and structured system of professional development activities, some of which were arranged by the school, and some were run by the District T.R. Center. Depending on where they took place, these professional development activities could be categorized into extra local, local, and school-based activities. Extra local professional development occurred outside the city; local professional development was organized by the local District T.R. Center; and of course, school-based professional development was held within the school building.

Extra local professional development

For many teachers at Merits School, their exposure to reform practices started from learning outside the city. Such learning opportunities were highlighted from time to time, when teachers talked about their experiences of the reform. Teachers had several opportunities to attend workshops, lesson observations, or textbook training within and outside the province. In addition, the school had two partner schools in Beijing. Every year, the school selected a number of mathematics teachers to spend one week there observing lessons. Since 2002, Teacher Rui had been to Beijing four times, Teacher Feng had been there twice, Teacher Hong had been to Beijing twice and to another two provinces once, and Teacher Zhang had been to Beijing and another city.

Extra local professional development appeared to expand teachers' professional horizons. It allowed them to have access to ideas and teaching, which might be particularly eye opening for teachers from less developed cities like the City. One unforgettable event happened to Teacher Zhang in November 2002, the second year of her teaching career. When the reform had just started, as one of the promising young teachers in the school, Teacher Zhang was chosen to go observe the province-level lesson competition that was being held in another city. The event lasted three days. Lessons in the competition were all well refined ones and delivered by expert teachers. Teacher Zhang took this learning opportunity seriously:

> I paid extreme attention to their instruction and observed very carefully. When I observed their lessons, I would put down as much as possible in my observation notebook, like every sentence they said, every question they asked, as well as students' answers. At that time, I didn't have the resources to tape-record the lessons then, so I even sketched the teachers in the margins of my notebook, detailing their facial expressions and manners. Anyway, I managed to memorize every teacher, and their teaching styles. After I got back, I modified some example lesson plans, and tried them in my own class. Indeed, it elevated my teaching.
>
> (Teacher Zhang, Interview #3)

To Teacher Zhang, it was "a very rewarding and eye-opening experience," which exposed her to the emerging reform thinking and practices. That experience set the tone for her active engagement in the curriculum reform.

Like Teacher Zhang, Teacher Rui spoke highly of those professional trips to Beijing. Teachers who gave demonstration lessons in the host schools were all outstanding teachers, and their lessons had been prepared and refined multiple times. Teacher Rui suspected that perhaps those teachers just honed one or two lessons over 1 year. She learned much from them. Particularly, she noted, she was better informed in terms of how to recognize and realize students' ownership of learning, integrate manipulatives into learning, and use hands-on activities properly. She put it this way:

> We really learned an awful lot from those lessons. Sometimes, maybe one delicate detail or one way of asking questions was enlightening... First, their students' status in learning was treated really highly...so

> their students enjoyed learning, absolutely liked learning, and kept high morale. Students asked very broad and excellent questions, and their language was truly elegant. From that, you could imagine, in their regular lessons, interactions between teachers and students were the same: rich and lovely. The second thing is that their instructional methods were much advanced. They used a lot of instructional manipulatives that kids liked playing with. In the meantime, teachers could still keep students engaged in learning the content…and students did not flee …because they integrated the content into manipulatives in the first place, for example, asking students challenging questions first. Then students would use the manipulatives in order to figure out the questions. All in all, we found that their ideas were much advanced. They could learn by playing. For example, in learning symmetry, they did not teach, but asked students to handcraft paper cutouts …Students indeed learned better when we used their methods in my teaching.
>
> (Teacher Rui, Interview #2)

Not every teacher had the opportunity for extra local professional development. Hence, after they returned to the school, teachers who went outside were required to hold a sharing seminar and to offer report lessons that were open to the whole school. Each of them was asked to prepare one lesson, reflectively incorporating what they had learned, to demonstrate to their colleagues. This follow-up event "not only helped spread the new ideas and but also helped one to internalize such learning in one's own teaching" (Teacher Feng, Interview #2).

The district's 2-4-8 Project

In 2003, the District T.R. Center put forward a new proposal named the 2-4-8 Project. According to the plan, every T.R. fellow was assigned to be in charge of one specific grade level and a number of schools in the district. He or she should identify and focus on two key schools, four focal T.R. groups, and eight backbone or promising teachers (hence the name "2-4-8 Project"). By training the eight teachers first and then capitalizing on these agents, the district expected to catalyze district-wide pedagogical transformation. Teacher Zhang and Teacher Feng were both involved in the project.

Centering on the core teachers, a variety of activities were organized from 2003 to 2005. In particular, one model was called *multiple teachers teach the same lesson*. Teacher Zhang explained the model in greater detail:

> We had a core team consisting of six or eight teachers from different schools. In the beginning of the semester, Fellow Han would plan out the focal themes that we were going to study over the semester, for example, topics like how to teach triangles, or fractions. After the event was formally started, usually in the second month, we would each prepare

> and teach lessons on the same topic. So, on Monday we would gather at School A and observe this teacher teaching for the whole morning, Tuesday at School B and observe another teacher, and Wednesday at School C. Finally we would convene at central schools and summarize the lesson. We did not simply observe but were required to give personal opinions on the lessons. After each observation, every teacher needed to comment on the strengthens and weaknesses of the lesson, in terms of how well the lesson was prepared, how the teacher carried out the instruction, or to what extent students grasped the content, and so on.
>
> (Teacher Zhang, Interview #3)

Teacher Zhang had benefited much from the Project. When the reform was in its heyday in 2003 and 2004, every week she went to four different places and observed four different teachers. Over 2 years, she observed about 260 lessons. She concluded, "Once we have observed and reviewed a lot, we become knowledgeable of which teacher has what issues on which topics, and we have greater capacity to plan lessons well to cater to the needs of students" (Teacher Zhang, Interview #3).

Teacher Feng and her Grade 3 colleague Teacher Min both took part in the 2-4-8 Project under the lead of another T.R. fellow. Teacher Feng described their prior experience of *multiple teachers teach the same lesson*:

> The lesson was on understanding 1,000, Grade 2 content. In total, it was taught four times by different teachers. The first teacher split his class into about ten groups, each consisting of five to six students. He gave each group 1,000 wooden rods, asked students to count them out and then report how they worked. Students had different ways to count. Some groups counted by ones. Some were cleverer, and when they had ten rods, they made one bundle. Then, for example, if a student had 150 rods, he would count by 1 ten, 2 tens, 3 tens, and so forth. Later on, some students found that ten bundles of tens made one hundred, so they made ten bundles of tens into one large bundle. The whole counting method of his lesson was good. But the problem with his teaching was that the rods were too small to hold, and they were spread all over the place. Some were even lost. It was quite messy. The second teacher replaced rods with straws. Furthermore, each group was given 100 straws instead of 1,000. The whole class added up to 1,000. He asked the groups to count on their own at first, and then he led the whole class to count together. But his counting method was somewhat chaotic. This time, he counted by hundreds, and that time, by ones. Then we found the second teacher's use of manipulatives was better than the first, but the first teacher's counting method was more systematic. So, we combined the strengths of both teachers and re-taught the lesson. The difficult point was to compose hundreds into one thousand. In order to crack the difficult point, the third teacher gathered all straws on the front teacher's desk and organized them by hundreds, tens, and ones. She led

> the class to count together. When having ten ones, they made one bundle of tens; when having ten bundles of tens, they made one large bundle of hundreds, until they had nine hundreds and nine tens. Continuing to count: 991, 992, ..., 999. This was the juncture. Adding one more, it was nine hundreds, nine tens, and ten ones. Ten ones make one ten, ten tens make one hundred, and eventually ten hundreds make one thousand. It was very visual for students to figure out how 1,000 was made.
>
> (Teacher Feng, Interview #2)

The lesson was then polished on the basis of the third teacher's instruction and taught the fourth time, towards which the whole team finally reached consensus. Approaching the end of the semester, the T.R. fellow called together all Grade 2 teachers in the whole district. Two *open lessons* (*gong kai ke*; *gong kai*=public, *ke*=lesson) were demonstrated at the event, one of which was the refined lesson on understanding 1,000.

Teacher Feng concluded that both novice and veteran teachers had benefited from the model in various respects. Novice teachers were given more opportunities to take responsibilities in public, had more exposure to varied instructional styles, and thus became mature and experienced more quickly. Similarly, veteran teachers had much to learn from young teachers who were more receptive to up-to-date pedagogies. From preparing, observing, and reflecting upon open lessons, teachers were able to derive a common body of content and pedagogical knowledge. "At the very minimum," Teacher Min stated, "we have thoroughly studied and grasped the text. To be honest, we can teach these lessons even with our eyes closed!" (Teacher Min, Interview #1).

School-based teaching research

Another professional development that teachers underwent was school-based teaching research. Schools in the City were required to institutionalize school-based teaching research pursuant to the regulations from higher authorities[1]. At Merits School, two major forms of school-based teaching research were organized: One was the grade-level weekly teaching research activity, and the other was the weeklong intensive teaching research event, *rotating to teach and rotating to observe* (*hu jiao lun ting*; *hu*=mutual, *jiao*=teach, *lun*=rotate, *ting*=listen or observe).

Grade-level weekly teaching research. Merits School reserved the last hour on Thursday afternoons for group-based learning and teaching research. On the occasion of teaching research, T.D. Wang and ten A.T.D.s were asked to participate in the groups that they took charge of. As required by the district, each grade-level T.R. group needed to put forward a plan in the beginning of the semester that detailed weekly research topics. During the weekly event, one key speaker was assigned to coordinate the discussion. Table 6.1 shows the weekly plan for Grade 1.

Table 6.1 Weekly research topics and key speakers at Merits School (Grade 1)

Date	*Research Topic*	*Key Speaker*
3.05	Orientation.	Zhang
3.12	Within 20 subtraction with decomposition.	Rui
3.19		Wu
3.26	Figure combination.	Yan
4.02	Knowing numbers within 100.	Tang
4.09	Reading and writing numbers.	Ying
4.16	Ordering and comparing numbers.	Zhang
4.23	Ones, tens and subtraction.	Rui
4.30	Using Chinese money.	Wu
5.07	Calculating money.	Yan
5.14	+/– within 100.	Tang
5.21	Tens +/– tens.	Ying
5.28	Two digits + one digit and tens.	Zhang
6.04	Knowing time and mini store.	Rui
6.11	Identifying patterns.	Wu
6.18	Statistics.	Yan
6.25	General review.	Tang

Grade-level weekly teaching research activities were run as follows. A.T.D. Zhou, who was also a Grade 5 mathematics teacher, visited the Grade 6 T.R. group. Teacher Su and another five colleagues were going to study the textbook and teacher's guide on the topic of solid prisms and cubes. It was their first meeting of the semester. Teacher Su, the grade-level team leader, anchored the discussion. Teacher Nan was in charge of keeping the minutes. Over the 35-minute meeting, they touched upon four issues: Experiencing solid prisms and cubes, and understanding their attributes; unfolding solid prisms and cubes; defining the length and the width; and calculating the surface area. On the first issue, they said:

SU: Make sure to have students look for solid prisms and cubes at home and bring them to the class. To introduce the topic, let them observe and experience these shapes first, for example, with their neighbors or in small groups, and then ask them what they have found. Now, let's review together the attributes of solid prisms and cubes. Here, there are two concepts, the edge and the vertex. The line along which two faces intersect is called an edge; the point at which three edges intersect is called a vertex. We have learned plane figures from a point, a line, to a surface. As for solid figures, we should teach backwards.

HONG: Ask students to identify faces first, and then edges and vertices.

SU: Exactly. They should be able to say that a solid prism has six faces, 12 edges, and eight vertices. The three edges meeting at the vertex are defined as the length, the width, and the height. That is to say, a cube is a special form of a solid prism. Why is it special? Its length, width, and height are equal. When only the length and the width are equal, then the opposite faces are squares, the other four faces are the same.

NAN: Is this a pattern that you found?

SU: It is fairly straightforward. If the adjacent two faces are squares, then the shape is a cube. From this, we can derive that at a maximum a [regular] solid prism has two square faces. On another matter, when we observe a cube or solid prism, how many faces can you see at most?

OTHER TEACHERS IN CHORUS: Three?

SU: But, there is a special case. If you place a solid prism, whose length is shorter than the distance between two eyes, close enough to the nose, you can observe four faces.

NAN: Then what should students answer in tests? Three or four? Take this exercise problem as an example, "Observing the solid prism from different angles, how many faces can you see at most?"

SU: This will not be tested. Answering four is also correct.

HONG: But the teacher's guide says three faces. I am curious why this question is given here? Doesn't it belong to the chapter of observing objects? We learned that content years ago. Why is it put here again? What is the purpose? I feel it is not closely connected to the themes of the unit.

SU: I think it is simply to provide more knowledge to students. The focus of the unit is to know the shapes, fully, globally and from multiple angles.

ZHOU: So, for regular sizes of cubes or solid prisms, we see three faces.

SU: In tests, we should still answer three faces.

HONG: In addition, a regular solid prism has at most eight equal edges…

(Observation in Grade 5 T.R. Group)

Group-based teaching research activities at Merits School took the form described above. At the beginning of the discussion, Teacher Su shared how he would handle this content. Further, he led his group members to go through the major concepts and the attributes of solid prisms and cubes. They defined edges, vertices, the length and so on. From there, teachers differentiated cubes from solid prisms, brainstormed possible special solid prisms (with two square faces), and made clear under which conditions a solid prism should become a cube. The focus on testing was evident in the discussion. Regarding how many faces one could see when observing a solid prism or cube, teacher Su introduced one special case to broaden his colleagues' knowledge. To keep consistent with the teacher's guide, teachers agreed on the right answer for tests.

Several teachers stated that discussions within the T.R. group had been particularly helpful for them to get familiar with the reformed curriculum in the first 2 years of the reform. Teacher Zhang agreed with that statement. For her, it had served as an important occasion to sit down with colleagues and to thoroughly examine specific content areas, which would otherwise leave one to figure out in isolation. Unfortunately, the structured grade-level teaching research had its downside. In their everyday work, many teachers did not have genuine discourses and collaborations. Frankly, Teacher Su confessed to me, "This is simply for show. These contents should have been discussed in our everyday work" (Pre-observation Conversation).

Lack of time appeared a legitimate reason for many teachers to explain why they did not work with peers. Teacher Tang calculated the many hours that she had to spend grading student work day after day: At least 3 hours per day. The task was so monotonous, and physically and mentally demanding that no teachers liked their job at all, not to mention the contrived teaching research. Teacher Zhang felt exhausted too:

> I am a mathematics teacher and classroom director in the classroom. In the T.R. group, I am the group leader. In the school, I am one associate teaching director. Too many distractions. How can I find the time to focus on perfecting my teaching? Besides, [because] our school is exemplary and a showcase in the district, if there are external activities, like inspections or visits, the district will bring them here. As you have seen, the province is said to be coming to inspect us on the 3-3-2-1-1 Project[2]. We have been dealing with that since the beginning of this semester, like preparing written materials, teaching students to practice the eurhythmics… We had to even put off our R.T.R.O. [rotating to teach and rotating to observe]. Normally, we should have finished that in late September. We wish we could have more time to focus on teaching and teaching research.
>
> (Teacher Zhang, Interview #3)

Echoing Teacher Zhang's comments, Teacher Hong also expressed her frustration. In addition to all the problems her colleagues faced, she brought up another issue: They had to spend a great deal of energy in coping with tedious and meaningless paperwork. Consequentially, teachers were distracted from participating fully in the weekly teaching research.

Rotating to teach and rotating to observe (R.T.R.O.). Merits School's intensive professional development event, R.T.R.O., played a significant role in experimenting with and popularizing the reform pedagogy. This professional development scheme had four steps: Lesson planning, instruction, lesson re-preparation, and re-instruction, similar to Lesson Study in Japan (Yoshida, 1999). Teachers first prepared lessons individually, coupled with spontaneous discussions with peers in the group. Following individual planning, a formal group-wide discussion took place. By capitalizing on collective knowledge, teachers reflected on their own lesson plans and re-prepared the lesson. As for instruction, there were two forms of activity: One form was that each teacher in the group taught the same lesson once with the rest observing; and the other form was that one teacher taught the same lesson multiple times for others to observe, critique, and revise the lesson. To conclude the cycle of R.T.R.O., teachers re-prepared and re-taught the lesson.

R.T.R.O. was used at Merits School for over three decades. It was generally held once every semester, for one week. Lessons demonstrated at the event were more reform-oriented and represent these teachers' more advanced, if not the highest, proficiencies, as differed from their regular day-to-day teaching. In the first 2 years of the reform, every teacher had been requested to give reform-based lessons. As a reflection of the district's policy change, R.T.R.O. emphasized different things too. Between 2003 and 2005, the focus of R.T.R.O. was modeling reform-oriented lessons that fulfilled the three-dimensional objectives (that is, knowledge, skills, and affective and attitudinal objectives) in instruction. Using computer technologies to aid instruction, promoting small group-based cooperative learning, and conducting hands-on activities were popular during that period of time. From 2005 onwards, the focus of R.T.R.O. shifted to collective lesson planning in order to have teachers know the curriculum better. R.T.R.O. created the venue for teachers to collaborate to some extent, and made it possible for them to observe, imitate, and learn from one another. Teacher Zhang put it this way:

> Basically we focused on one unit or one topic over the week. We would examine the standards, dig into the curriculum, and analyze the text and students together…After having taught the lesson for the first time, we would come back together and have the instructor explain what she had done well and what she had not, and then we would tackle the issues as a group. Following that, we would plan the lesson one more time. Sometimes, six of us would teach the same topic several times. This teacher might have these problems in her or her instruction, and the second might encounter other problems, but after all six people were done, there were very few problems at all. We got some refined research lessons. Certainly our teaching quality was improved.
>
> (Teacher Zhang, Interview #3)

Summary

Teachers' instructional methods at Merits School had changed to a certain extent: They used small groups more frequently, gave more hands-on activities, and were more attentive to students' input after the reform. To make those changes happen, varied professional development schemes were designed and institutionalized at the school and in the local district, and they played an important role in facilitating teachers' instructional change.

Notes

1 For instance, the C.E.B. put forth in its Opinions on the Administration of Teaching in Elementary and Secondary Schools in the City, "Schools should establish the following institutions: (1) the weekly teaching research institution, (2) the regular

meeting of T.R. group leaders, (3) the institution of reporting on, appraising, and rewarding achievements of teaching research, (4) the institution of teachers' attendance at group teaching research activities at the city, county (district) and school level, and (5) the evaluation institution of teaching research" (C.E.B., 2008, p.9).

2 The 3-3-2-1-1 Project was made up of:

Three classes: The physical education and health class, music class, and art class.
Three exercises: The morning exercise, the between-lessons exercise, and the eye protection exercise.
Two activities: The sports activity, and the science, technological and cultural activity.
One event: The track and field event.
One festival: The arts festival.

It was a well-intentioned initiative to improve students' heath, proposed by the Province Education Bureau in 2001.

7 Pioneer School embraced the reform

The curriculum change at Pioneer School was undertaken in a systematic fashion. With the support of the City T.R. Office, the school began first with a reform-minded experiment, the Innovative Elementary Mathematics Education (I.E.M.E.) Experiment, in late 1999. The fundamental ideas informing the I.E.M.E. Experiment were essentially consistent with those that guided the national mathematics curriculum reform. The I.E.M.E. Experiment paved the way for the school to embrace the national reform in 2002.

Thanks to its reform-oriented culture, the administrators and teachers at Pioneer School were collaborative in carrying out the I.E.M.E. Experiment and the subsequent reform. The school, together with the City T.R. Office, provided teachers with varied school-based and external professional development opportunities. A lot of change had occurred.

However, Pioneer School was demoted from being managed directly by the C.E.B. to being managed by the D.E.B. in 2006. The school's status change led administrators to believe that they experienced more unfavorable directives from the district. It was believed that intrusion from higher authorities substantially affected the school's autonomy to continue with the reform. Notably, throughout the process of the reform, parents were marginally involved.

The pre-reform experiment

The rationale for change

> The 21st century is a century full of hopes and challenges. It is a time characterized by knowledge economy. Inevitably, global competitions will become increasingly intense. It is necessary that we cultivate highly qualified talents who are innovative and creative. Critically examining the status quo of mathematics education at Pioneer School in the past century, it is not hard to find that mathematics teaching and learning in our school had long been textbook-bounded, classroom-confined, and teacher-centered. Owing to traditional thoughts of education,

> instructional modes had stuck to lecturing and indoctrination. We had relied on one test paper to assess learning and one singular approach to evaluating students. Students' subjective activity, developmental multiplicity, and variation had been overlooked, and their creativity had been inhibited...
>
> (Pioneer School, 2003a, p.1)

Out of the above considerations, administrators at Pioneer School, together with the City T.R. Office, intended to embark on a local experiment in elementary mathematics education, the I.E.M.E. Experiment in late 1999. The I.E.M.E. Experiment was heavily inquiry-oriented and congruent with the ideas of the formal national mathematics curriculum reform that would be adopted in the district 3 years later. In a sense, Pioneer School had already begun the curriculum reform with the I.E.M.E. Experiment ahead of all other schools in the district.

A brief depiction of the bigger picture behind the I.E.M.E. Experiment is due here. Approaching the dawn of the 21st century, calls to transform its basic education turned stronger and louder in China. In early 1999, the central government promulgated the *Decision to Strengthen the Educational Reform and to Foster Full-scale Qualities-oriented Education* (M.O.E., 1999). Echoing that call, China's M.O.E. began developing curriculum standards in all elementary subject areas. The task force on the development of mathematics curriculum standards was chaired by one professor from Beijing Normal University, who had previously been educated in the U.S.

Nearly at the same time, a number of schools across the country were experimenting with a new series of mathematics textbooks, *The 9-Year Compulsory Education Curricula-Mathematics* (Beijing Normal University, 1999), better known as *New Mathematics*. The series of books embodies many reform ideas that were later made present in the N.M.C. Standards (2001). For instance, the central tenets of the curriculum were: (1) Closely connecting mathematics with real life; (2) firmly assuring students' subjective status in mathematics learning; and c) exploring and establishing inquiry-based learning approaches, and cultivating students' innovative abilities. *New Mathematics* was regarded as "the most advanced curriculum that was designed by drawing on the best thoughts on mathematics education in Western countries" (Teacher Me, Interview #2), and it was considered avant-garde in form, rigorous in mathematics, rich in information, and inquiry-based in pedagogy. It was unlike the traditional curriculum that was rigid and dull, and generally left teachers with little room to demonstrate their own thinking. In contrast, *New Mathematics* could "provide teachers with flexibility...if you are a highly able teacher, you can expand far beyond the textbooks, and if you are a mediocre teacher, it is also fine to limit to the scope of the textbooks" (Teacher Me, Interview #2). The school decided to use *New Mathematics* in the I.E.M.E. Experiment.

Preparing for the IEME Experiment

Before 2006, Pioneer School had been directly led by the C.E.B., even though administratively it was attached to the D.E.B. The school's higher status than that of other schools in the district had thus granted it with greater autonomy. As Teacher Me commented, over its long history, Pioneer School had been the flagship school in the City, always first to try out new educational ideas. The C.E.B. desired to have at least one model school in the City. Thanks to the school's reform-oriented tradition, both the administrators and teachers had been habitually receptive to educational innovations. From the late 1970s to early 2000s, as Teacher Me recollected, the school had undertaken at least three rounds of mathematics curriculum change:

> In my earlier career, we were involved in learning from Xinlan Ma, one of the best known mathematics teachers in our country. Particularly, she had a system of approaches to solving word problems. We also experimented with the *three calculations*, namely, rapid calculation, abacus-aided calculation, and mental calculation. Another experiment was called *developing elementary mathematics*, which lasted 5 years.
>
> (Teacher Me, Interview #1)

Sensing the pedagogical shift in mathematics education, Fellow Xu from the City T.R. Office and Principal Bao, the former principal, concurred to keep up with the trend. On June 4, 1999, Fellow Xu, Principal Bao, V.P. Yang, and T.D. Zhi, held the first meeting in the school. They agreed that the "textbook-bounded, classroom-confined, and teacher-centered" (Pioneer School, 2003a, p.1) reality of mathematics education at Pioneer School should be changed, and new curriculum materials should be implemented. The school proposed to adopt *New Mathematics* in two Grade 1 classes in the coming fall semester. T.D. Zhi described *New Mathematics* in this way:

> It is really good, particularly conducive to freeing students' thinking and cultivating their creativity. The curriculum does not tell you exactly what to do, but offers teachers and students a large degree of freedom. For instance, what is the unit of area? It does directly tell you what 1 square centimeter is but asks students to explore the whole process of how this knowledge comes into being. It suggests the surface of a person's nail is about 1 square centimeter and the palm of one's hand is about 1 square decimeter. Of course, the drawback of the curriculum is that knowledge is not so systematically organized [as the traditional curriculum]. Like the above case, it should finally give the clear definition of the unit of area: The area of a 1-decimeter square is 1 square decimeter. But the text does not make that knowledge explicit.
>
> (T.D. Zhi, Interview #2)

Shortly after the meeting, the school formed the I.E.M.E. Experiment Steering Committee, and all 11 school administrators were on board. V.P. Yang, also a national backbone teacher in Chinese language, and T.D. Zhi, a provincial backbone teacher in mathematics, both sat on the Steering Committee. Four experienced teachers were singled out, including Teacher Me, a national backbone teacher, Teacher Chun, an A.T.D. in mathematics and provincial backbone teacher, Teacher Jun and Teacher Xiang, both provincial backbone teachers. Particularly, Teacher Me was the central figure among the four. She had spearheaded many educational experiments in the school before and was highly respected for her dedication and excellence in teaching. At the first team meeting on June 20, 1999, Principal Bao decided that they would support the I.E.M.E. Experiment teachers to learn how to use the new mathematics curriculum in Tianjin in the summer.

In July of that year, Fellow Xu led Teacher Xiang and another two experiment teachers to visit one elementary school in Tianjin, a province-level city in China, since that school had been beginning to use *New Mathematics*. Over the 3-day event, they observed Grade 1 mathematics lessons during which small group-based cooperative learning was showcased. In addition, those visitors collectively analyzed the textbook, *New Mathematics*, and engaged in discussions on special features of the textbook and how teachers should use the N.M.C. The editor of *New Mathematics* also gave a presentation that critically examined the lasting problems latent in China's basic mathematics education, and compared it with mathematics education in developed Western countries. Once more, the imperative to retooling the country's mathematics curriculum was made crystal clear. Back home, those three teachers would serve as the primary agents to catalyze the school-wide curriculum change.

The fall semester saw the initiation of the I.E.M.E. Experiment at Pioneer School. On September 10, 1999, Principal Bao, T.D. Zhi, Teacher Me, and the teachers who visited Tianjin got together. Teacher Xiang, and Teacher Chun, A.T.D. in mathematics, shared what they had learned from that professional exchange experience. They discussed extensively the possibility of using the reform pedagogy to teach mathematics at Pioneer School and began formulating the I.E.M.E. Experiment plan.

Following that meeting, Fellow Xu, three school administrators, and the I.E.M.E. Experiment teachers convened again on September 26, 1999. Fellow Xu stressed the importance of innovating mathematics education, with Principal Bao reiterating its feasibility to implement the I.E.M.E. Experiment. Principal Bao was delighted at the promising prospect of the Experiment and decided to provide as many resources as possible to support the Experiment endeavor at the school. It was promised that outstanding experiment teachers would gain priority for promotion or awards. Further, T.D. Zhi introduced the overall experiment plan for the coming 4 years, from 1999 to 2003. It said that from September 1999 to August 2003, the I.E.M.E. Experiment would be implemented gradually throughout the school. It requested that teachers' lesson preparation, instruction, homework design, and assessment of learning be fully innovated. In order to fulfill those goals, it directed:

(1) Maximizing the leadership of the school's renowned teachers (*ming shi*; *ming*=famous, *shi*=teacher), skilled teachers (*neng shou*; *neng*=able, *shou*=hand), and backbone teachers (*gu gan*), assigning them with a heavier load, having them teach at least one modeling lesson per semester, and helping at least one young teacher.
(2) Training teachers thoroughly by inviting external experts to give lectures, watching videos and so on.
(3) Improving experiment teachers' professional skills by carrying out teaching research activities on special topics, having one research lesson each month, and each semester having everyone give at least one research or demonstration lesson.
(4) Constructing innovative classroom instructional modalities.
(5) Formulating dynamic, scientific and innovative assessment approaches and giving each student one comprehensive evaluation on his or her learning performance per semester.
(6) Taking one lesson (1 or 2 hours) each week as the mathematics activity time in order to connect organically mathematics knowledge with life practices.
(7) Analyzing two cases on the mathematical activities that have been carried out.
(8) Having students design one mathematics handwritten poster every month in order to widen students' scope of knowledge and stimulate their interest in learning mathematics. (Pioneer School, 2003a, p.57)

Systematizing teaching research was also emphasized so as to have teachers frequently communicate with one another, overcome potential hurdles, and promote teacher learning. The school planned to engage teachers in a variety of professional development activities within and outside of the school. They planned to have all experiment teachers and administrators meet at least once a month. Participants would present to their colleagues innovated lessons and share their learning, when Fellow Xu would be invited to critique their teaching. Additionally, teachers were expected to write one research paper on a special topic, collect two teaching cases, keep three journals, and compile four issues of reform communication letters per semester.

General principles of the I.E.M.E. Experiment

On November 6, 1999, Principal Bao held a meeting and formally announced to the parents of those two Grade 1 classes that the school had started the I.E.M.E. Experiment. It was the only attempt that the school made to inform parents of curriculum change throughout the I.E.M.E. Experiment and the subsequent reform. The I.E.M.E. Experiment with its five-step teaching was first implemented in Teacher Me, Teacher Xiang, and Teacher Chun's classrooms. To guide the I.E.M.E. Experiment, as V.P. Yang noted, four general principles were put forward by the Committee (Pioneer School, 2003a, pp.2–4). First, teaching should be democratic. Teachers were expected to create a harmonious, safe, democratic, and equal classroom, within which students could think freely and dare to question. Second, students should actively acquire knowledge via hands-on operations and inquiries, which was regarded as the core tenet of the I.E.M.E. Experiment. Third, teachers were asked to respect students, consciously cultivate their qualities, and encourage them to pursue their own goals and to develop unique learning styles, interests, hobbies, and specialties. Last, teachers should make use of all kinds of approaches to stimulate learning so that students could experience the joy of success. Those principles represented the efforts to change a teacher-centered culture to a student-centered one.

By observing these principles, the I.E.M.E. Experiment strove to promote inquiry-based methods in mathematics instruction. For this purpose, a general flow of instruction was proposed by the reform team. According to this flow, a lesson should run from *creating the situation*, to *inquiring by self*, to *cooperating and communicating*, to *assessing diversely*, and to *summarizing and providing feedback*. The idea of *creating the situation* was:

> At the beginning of the lesson, [teachers] need to produce a pleasant classroom environment by means of storytelling, playing games, conversing, and the like, so that students can start learning in a good mood. [Teachers] may also utilize students' prior experiences or things that are close to their lives to create learning situations and excite students' curiosity. In this way, students' desire to learn can be piqued.
>
> (Pioneer School, 2003a, p.10)

It was also important for teachers to allow students to explore on their own and to dare to question during the process of learning new knowledge. Teachers were asked to encourage students to ask more why's, and "questions that can lead students to observe, think, inquire, and imagine" (Pioneer School, 2003a, p.11). Thus, it was expected that teachers should change their communication styles by using more encouraging language and giving students enough time and space to engage in inquiries. By so doing, teachers could cultivate students' reasoning, and be able to guide students to conduct learning in a systematic order, that is, from the simple to the complex, from the easy to the difficult, and from the superficial to the deep.

Furthermore, it stressed the importance of having students actively cooperate and communicate with one another in the process of inquiring new knowledge. Hands-on activities were placed at the core of inquiry-based learning, during which students could engage in hands-on practices, discussions, questioning, and reasoning and the like. Thus:

> Teachers should free students, enable them to manually practice and operate, and create opportunities for them to conduct open discussions, voice viewpoints, and learn from each other. Also, teachers should encourage students to generate fresh ideas, to look at an issue from multiple angles, and to dare to challenge the textbook and teachers. In this way, students' thinking can be developed, and their creativity be nurtured. Besides, students can learn to accept and appreciate others, mutually improve, and develop a cooperative ethos.
>
> (Pioneer School, 2003a, p.11)

As for assessment and periodical summarization, the school noted that constant assessment could provide timely feedback to teachers as well as students in order to promote student learning effectively. V.P. Yang claimed that it was essential that the I.E.M.E. Experiment be accompanied by the retooling of the test grade-oriented assessment system into one that was "holistic, specific, objective, reasonable" (Pioneer School, 2003a, p.12), and should take into consideration children's varied abilities. Otherwise, no substantive outcomes could be achieved or sustained.

In response to this need, the school made an effort to reshape the assessment system. The old system was said to have long been biased because of focusing too much on teaching and knowledge and less on learning and skills, and focusing more on results than the process, the required answers than independent thoughts, and administrators' evaluation than assessment by teachers and students themselves. Three features characterized the innovative assessment system. First, it emphasized that assessment should focus on multiple aspects of learning. Not only should students be tested in terms of how well they memorize the knowledge and how adeptly they solve problems, but also how students actually make use of what they have learned in tackling real life problems. Assessment should be conducive to the cultivation of students' hands-on abilities. Paper-pencil tests should increase the diversity of the contents tested and the flexibility of answers. Second, it specified that the process of assessment should be dynamic, meaning that assessment should pay due attention to both the products and the process of learning. Teachers were requested to give positive comments on student work. Last, the school attempted to employ diverse approaches to assessing student learning. For instance, oral tests, paper-pencil tests, and hands-on operations were all supposed to be an integral part of the assessment system. A dynamic and innovative assessment system was formed in order to assess students globally and develop their innovative abilities, as shown in Table 7.1.

Table 7.1 The system of academic assessment at Pioneer School

First-level criteria	*Second-level criteria*	*Third-level criteria*	*Descriptors*	*LoP*		
				E	*G*	*P*
Classroom performance	Learning approaches with positive attitudes	Affection	(1) Being fond of learning mathematics & full of curiosity. (2) Hard-working, confident, and motivated.			
		Learning approaches	(1) Taking the initiative to learn and daring to pose unique questions. (2) Being willing to cooperate and communicate with peers. (3) Being willing to listen to other students and teachers attentively. (4) Being able to use mathematics knowledge to solve problems in life. (5) Having unique solutions.			
	Multiple intelligences	Observing	Being able to observe precisely and identify mathematical issues.			
		Operating	Being able to use manipulatives and objects to learn relevant knowledge, and having strong abilities to execute operations.			
		Expressing	Being able to express logically one's own understanding and the process of problem-solving.			
		Creating	Being able to use different approaches to thinking and solving problems.			
Knowledge and skills	Student work	In-class and homework	Writing neatly, finishing on time, using methods flexibly and correctly, and correcting errors immediately.			
	Unit or periodical test	Items	Basics Calculation Problem-solving Operation Practice			
		Grades				
End-semester evaluation						

Note: LoP=level of performance, E=excellent, G=good, P=poor.

Translated by the author from Pioneer School (1999).

Learning to implement the IEME Experiment

Collective lesson study became an integral part of teachers' learning to teach in the new instructional model. Under the coordination of T.D. Zhi, the school held a 2-day lesson demonstration event on March 17–18, 2000. All experiment teachers in the school took part in the activity. Teacher Me, Teacher Xiang, and Teacher Chun all gave model lessons observing the four principles. Fellow Xu was invited to critique and help improve their teaching. The outcomes of those research lessons were mixed. On the one hand, teachers' instruction showed substantive change. According to T.D. Zhi:

> We used a lot of small groups to demonstrate what cooperative learning entailed. Over a 40-minute lesson, teachers were busy organizing groups, distributing materials and learning aids, giving directions, and guiding discussions. It was so different from our traditional view of a mathematics lesson. Before, we would basically listen to the teacher who had the script of the lesson, and students would not have the chance to bring their views and activity into learning. Even though traditional instruction might also pose a lot of questions to students and seemingly have them reason, actually those questions were prescribed ones. Teachers determined beforehand what to ask, and students cooperated to answer the questions in the teachers' favor.
>
> (T.D. Zhi, Interview #1)

On the other hand, these lessons revealed administrators' and teachers' lack of understanding of small group-based cooperative learning. T.D. Zhi remembered that some doubts were:

> What was going on? The classrooms were noisy and seemed out of control, students were talking to one another, playing with objects, and moving around, and teachers stood aside and were not supposed to give explicit answers or even talk. How practical could small group-based cooperative learning be? Could it satisfy the need for instructional efficiency and effectiveness?
>
> (T.D. Zhi, Interview #1)

Such questions were typical and persisted in the beginning of the I.E.M.E. Experiment. Because of limited and mechanical understanding of the reform ideas, school people could not find the balance between instructing and allowing students to explore and apply on their own. As V.P. Yang pointed out:

> The reform thinking says, "Students are the owners of learning, and teachers are participants." Then, teachers were totally lost and they didn't even know how to teach any longer. Students were active. The classroom became chaos. No meaningful knowledge could be learned. The whole 40 minutes was wasted for an impressive but meaningless show. So, we asked

ourselves, "What is the role of teachers in the N.M.C.? What is the relationship between teachers and students? To what end is having students work in groups and do hands-on projects?"

(V.P. Yang, Interview #1)

Faced with these issues, the I.E.M.E. Experiment proceeded and tackled them through further trial and learning. The issues that challenged the I.E.M.E. Experiment in the beginning were gradually figured out. The misunderstanding about teachers' passive role in small group-based cooperative learning was dismissed, and teaching staff came to realize that teachers need not give up their authority of being more knowledgeable people than students and they should actively observe, probe, and lead students to explore key mathematics concepts. Lesson opening was another example. As Teacher Me recounted, not every lesson should start with a story or activity: Students in the lower bands would enjoy listening to a story, but more advanced students would need to be challenged by the beauty of mathematics.

In subsequent years, the I.E.M.E. Experiment expanded to all five Grade 1 classes starting in the fall of 2000 and seven more in the fall of 2001. Six more teachers were involved in the I.E.M.E. Experiment. Teacher Wen, and another three teachers had all taught mathematics for more than 20 years in the school, while Teacher Quan and the other young teacher were in their third and second years of teaching. T.D. Zhi expected that inexperienced teachers could be rapidly developed into mature and capable professionals under the influence of experienced teachers, while veteran teachers could also be inspired by the innovative spirit of young teachers.

In order to facilitate teachers' learning and to provide the venue for collegial exchange, the school regularly arranged collective lesson study activities. From March 2000 to the final evaluation of the I.E.M.E. Experiment in October 2003, at least 11 lesson studies were held for all experiment teachers so that they could be engaged in in-depth critique and discussion of the reform-minded lessons (see Table A1 in Appendix 1 for the major experiment events that took place at Pioneer School). Such learning opportunities had "effectively transformed teachers' attitudes towards teaching, learning and students" (T.D. Zhi, Interview #1).

One example was that teachers had consistently encouraged students to write mathematics journals. In the past, teachers had relied on test papers, remote from students' everyday experience, to gauge learning. Writing mathematics journals had been a frequently used approach at Pioneer School. For the purpose of illustration, Table 7.2 displays one Grade 2 student's mathematics journal.

The above example was a typical entry in a mathematics journal. The student consciously inquired into his or her environment and discovered the mathematics present in everyday life, such as, time, money, and measurement. The means of writing a journal was an effective approach to both assessing student learning and helping students build up the connection between mathematics and their lives.

Table 7.2 A mathematics journal from a Grade 2 student at Pioneer School

Sunday October 10, 2002
Everyday accumulation
Today, the homework is again especially interesting – investigating some approximate and exact numbers in everyday life.
We have found many such materials. The approximate numbers we found are: A person needs to consume approximately 1 kilogram of water and to sleep 7 hours per day; the ordinary abacus has about 15 places. The exact numbers have: The clock surface has 12 long tick marks and 60 short tick marks, 1 minute is equal to 60 seconds, 1 hour is equal to 60 minutes, and 1 hour is equal to 3,600 seconds; one meter is equal to 10 decimeters, 1 centimeter is equal to 10 millimeters, 1 meter is equal to 100 centimeters, 1 meter is equal to 1,000 millimeters, and 1 decimeter is equal to 100 millimeters; 1 yuan is equal to 10 jiao, 1 jiao is equal to 10 cents, and 1 yuan is equal to 100 cents. The above is the result of my investigation.
As long as we observe carefully, we will find more approximate and exact numbers in everyday life. We will not cite every case here.

Note: Translated from Pioneer School (2003a, p.207).

The emergence of the N.M.C. Reform

As the school was pioneering to innovate mathematics teaching and learning, China was preparing to launch an ambitious curriculum reform that concerned all subject areas and grade levels from elementary to high school. V.P. Yang remembered how she felt when she first heard about the reform in late 1999:

> I was very excited and much relieved from the bottom of my heart, as if we had finally seen the light in the dark. As frontier teachers, we all looked forward to the day. The proposed reform was ambitious in its scale and depth. Indeed, there were many wonderful groundbreaking ideas in the reform initiatives. The N.M.C. not only paid due attention to the traditional core of learning – knowledge – but also valued the development of practical skills, the cultivation of positive attitudes towards learning.
>
> (V.P. Yang, Interview #1)

As for elementary mathematics, M.O.E. made a substantive move towards the national mathematics curriculum reform in early 2000: The first draft of the N.M.C. Standards was released by M.O.E. to the public for feedback. In late April 2000, Pioneer School attended a one-day training on the draft of the N.M.C. Standards, organized by the City T.R. Office and aimed at city-level and above backbone teachers. From 2000, the I.E.M.E. Experiment was influenced by the training under the draft Standards.

In the meantime, M.O.E. started training national backbone teachers across the country who would later serve as reform leaders in their own regions in the upcoming grand reform. As V.P. Yang described:

> From September 2000 to October 2001, Teacher Me pursued part-time advanced studies at one major normal university in Beijing. In the summer of 2001, she was sent to Nanjing to learn things about the N.M.C. reform. During the training events, many famous professors and scholars came to give talks about the reform and helped us make sense of the general reform spirit and specific curriculum standards. It was exactly brainwashing, in a good way: We could get rid of the traditional teacher-directed mentality and develop inquiry-oriented cooperative thinking.
>
> (V.P. Yang, Interview #1)

In early 2002, the City officially launched the national curriculum reform. Following in the City's footsteps, two separate committees were formed at Pioneer School. One committee was called the Steering Committee for the Curriculum Reform, consisting of the same 11 school administrators as the I.E.M.E. Experiment Steering Committee. The other committee was the Consulting Committee for the Curriculum Reform, made up of key teachers in Chinese language, mathematics, and English language at the school. At that point, the I.E.M.E. Experiment at Pioneer School essentially merged into the reform.

With regard to the mathematics curriculum reform, the school capitalized predominantly on what it had achieved over the 3-year experiment. The Steering Committee drew up a general blueprint which illustrated the purpose and goals of the reform, and the procedures and strategies to actualize it. Actually, the reform plan had no essential differences from the former experiment plan. Several major goals were reiterated, that is, strengthening reform-minded professional development for teachers, innovating inquiry-based pedagogy, and enhancing students' abilities to problem solve and process information, reforming approaches to assessment, developing school-based curriculum, and promoting school-based teaching research and nurturing reflective and research-minded teachers.

From September 2002 onward, all Grade 1 classes, including the 5-year track and the 6-year track, started implementing the curriculum reform. In 2003, the school stopped using *New Mathematics*, since the district decided to adopt one common curriculum written and published by Jiangsu province, which was then replaced in 2006 by the curriculum published by one major textbook publisher, People's Education Press, in Beijing.

The reorganization of the school in 2006

The historical event that had a significant impact on the curriculum reform occurred at Pioneer School in 2006, when the C.E.B. restructured the local school system and Pioneer School was demoted from the level where they were directly managed by the C.E.B. to the level where they were directly managed by the D.E.B. That action ended the school's status as superior to the rest of elementary schools in the district that it had enjoyed for the previous 40 years.

Consequentially, the official status of the school's Principal and V.P.'s was lowered. Principal Bao and V.P. Yang had been on a par with the heads of the D.E.B., but from 2006 onward, they became subordinate to the D.E.B. officials. One of the most obvious impacts was that the school lost most the freedom it had once enjoyed, which distracted the school away from teaching and inevitably hampered the school's reform efforts. V.P. Yang lamented:

> There are far too many orders and the interference [with the school's own business] from outside of the school. Today they demand this and tomorrow that. Today they want this report and tomorrow that. We feel that we all are physically and mentally exhausted and can hardly cope with the demands. We are deprived of our time for learning and self-improvement. It is the same situation for ordinary teachers. Like this one, the Teacher's Professional Learning Notes, it is mandated by the D.E.B. that teachers must turn in reading notes that show evidence of learning. However, it simply takes away teachers' time that they could have used for their real inner learning needs... Few of those higher-ups are truly knowledgeable of schools' business and every newly elected administration will come up with something new, which, however, stands in the way of the school's operation.
>
> (V.P. Yang, Interview #1)

The directives from external educational authorities were judged to be excessive, arbitrary, and interfering with the school's genuine needs. Particularly, some local higher authorities' ignorance of the N.M.C. rendered implementation of the reform ineffective. V.P. Yang put it this way:

> The main heads in the D.E.B. and the local T.R. Center don't know what the essence of education is. Their understanding of the N.M.C. is at best superficial and ceremonial. Think about that yourself: How can these authoritative offices guide us, while they themselves cannot meaningfully grasp the curriculum standards? You went to the training session on new curricula by the District T.R. Center last week, right? [I nod.] What did you see? It was supposed to be a 3-hour session, but that fellow got it done in less than an hour. Such activities are barely meaningful. Teachers cannot learn much.
>
> (V.P. Yang, Interview #1)

Echoing V.P. Yang, Mr. Huang, the current Principal of Pioneer School, shared similar sentiments regarding the lack of autonomy. In post for less than 1 year, he experienced many disruptions from some higher-ups in running the school. Principal Huang said:

> Every head up there wants to do something different to polish their achievement. Some of their notions are simply out of touch with the school's reality, too out-of-date or random. Like this, Memo of Tutoring

> Students with Learning Difficulties, it has good intentions but is not practical. Teachers have no time to really visit students' homes, so most content recorded in it is made up.
>
> (Principal Huang, Conversation in the Playground)

Apart from external intrusions, more problematically, the school was now subject to evaluation by the D.E.B. and the District T.R. Center. As mentioned in Chapter 4, in 2006, the District T.R. Center restored the district-wide Uniform Examination, which took place in the first or second week of July. All students in the district would take the same standardized test written by one of the T.R. fellows. Grade 3 and Grade 6 students' test papers would be graded at the District T.R. Center and average grades would be calculated. Even though the T.R. Office declared that no schools' ranking information would be published, all schools knew where they stood compared to their peers. Pioneer School was no longer exceptional, although it had had the privilege to design and administer its own tests. The N.M.C. reform was thereby offset to a significant degree:

> These recent few years, we think the reform is backpedaling: The measure that the District T.R. Center takes to evaluate schools is the baton. All schools keep their eyes on it. Once the District T.R. Center demands test grades and ranks schools, all schools turn to focus on exams.
>
> (V.P. Yang, Interview #1)

In that sense, Principal Huang commented, "the N.M.C. reform [of the City] has essentially failed" (Interview #1), and most schools returned to the old track that education was about getting good test grades.

Striving to make a difference

In spite of the increasingly unpleasant milieu, Pioneer School strived to keep its unique features, that is, having relatively lower school workload and offering a richer curriculum compared to other schools in the City. As Teacher Me said, liberal culture had taken root in the school over the previous 40 years. While making sure that the school would not perform poorly in the district tests, Principal Huang did not attempt to orient the school towards heavy drills and prolonged hours of exercising.

Responding to one teachers' persistent complaints, the large class size, Principal Huang assured teachers that he would strictly abide by the official class size, a maximum of 48 students per classroom in every school, which had been stipulated by the C.E.B. long before and stressed over and over again, but never enforced at the school. Indeed, at the beginning of the school year, Principal Huang fulfilled his promise: None of the six incoming Grade 1 classes exceeded the upper limit of 48 students. Probably, that was the most that the school could do for the good of teachers and students.

Table 7.3 Teacher Me's Grade 1 curriculum schedule

Curriculum schedule					
Day No.	*Mon*	*Tue*	*Wed*	*Thurs*	*Fri*
1	Math	Chinese	Math	Chinese	Math
2	Chinese	Math	Chinese	English[a]	Chinese
3	Morning exercise	Morning exercise	Morning exercise	Morning exercise	Morning exercise
4	P.E.	Science	Art	Reading	Chinese
5	English[a]	Music	Moral education		Art
6	Science	Chinese	English[a]	Music	Moral education
7	Moral education	P.E.	Chinese	School-based curriculum	Comprehensive class
8	Afternoon recess	[b]	Afternoon recess	Afternoon recess	Afternoon recess
9	Class meeting	[b]	Handcraft	P.E.	Art

Note: Morning class runs from 8:00 am to 12:00 noon in summertime. School finished after the fifth lesson and restarts at 2:30 pm.

[a] Indicates that the three periods of English are given to mathematics instead.
[b] Indicates that students are dismissed after two afternoon lessons and the remaining time is reserved for weekly school-level collective learning of teachers.

The school managed to follow the national curriculum schedule to a greater extent. All para-subjects, such as, music, art, and physical education and the like, were offered to the last week of the semester. Table 7.3 illustrates the school's actual curriculum schedule. The school had seven 40-minute periods of mathematics lessons, which, according to T.D. Zhi, was a more realistic amount of time, the least needed to ensure students' mathematics performance.

Summary

Being reform-minded in nature, Pioneer School had been in the leading position to proactively initiate and implement the N.M.C. Standards. The school had started it with the 3-year I.E.M.E. Experiment and then merged with the nationwide curriculum reform. School administrators, the past principal, V.P., and T.D. were found to be genuinely supportive of the reform and had managed to provide teachers with opportunities for professional development within and outside the school. Yet, the new mathematics curriculum was not fully implemented. Partly, the demotion of the school's official status in 2006 had put the school in a more competitive exam-oriented environment with excessive external directives from the district education administration. Even so, the school strived to maintain its relatively humanistic culture catering to the multiplicity of students' needs.

8 Teaching evaluation at Pioneer School in response to the reform

> In order to further enhance teachers' capacities, stimulate teachers' enthusiasm and passion, and formulate an evaluation system that meets the needs of the N.M.C. reform and the development of teaching, [the school] particularly stipulates the following general principles and detailed approaches to evaluating teaching. Teaching evaluation lays emphasis on stimulating teachers' inner need and motivation for continuous development so as to enable every teacher to best utilize each one's unique advantages and to have teachers analyze and reflect upon their own teaching behaviors. In the end, teachers can meet the standards that the N.M.C. sets for teachers' qualities and professional calibers.
>
> (Pioneer School, 2003b, p.1)

School administrators at Pioneer School restructured its teaching evaluation system in line with the principles of the N.M.C. Reflecting the call to diversify the approaches to teaching evaluation, they attempted to draw on multiple sources of information about a teacher's professional behaviors. Both teachers' teaching performance and ethics were taken into account. The school administrators, including principals and T.D.s, grade-level team leaders, and teachers themselves could contribute 70%, 20%, and 10%, respectively, to the evaluation input. A variety of measures, lesson observation, surveying, and evaluations of student work, teachers' lesson plans, and periodical reports, and the like, were taken to collect data on teachers' performance.

To paint a general picture of the system, Table 8.1, the Teaching Evaluation Spreadsheet of Pioneer School, presents the end-of-school year evaluation results of nine mathematics teachers. The system contains six main categories: Teaching norms, behavior, school-based teaching research (S.B.T.R.), meeting attendance and paperwork (M.A.P.), parent review, and quality of teaching. Several main categories also consist of further delineated subcategories; for instance, under the category of teaching norms are lesson planning, instructing, and student work.

Table 8.1 The teaching evaluation spreadsheet at Pioneer School

Name	*Teaching Norms (25 pts)*			*Behavior (30 pts)*		*S.B.T.R. (10 pts)*	*M.A.P. (5 pts)*		*Parent review (5 pts)*	*Quality of teaching (25 pts)*			*Total*
	L.P.	*I*	*S.W.*	*W.L.*	*Attendance*		*Meeting*	*P.W.*		*Unit*	*Final*[a]	*Contest*[b]	
Hong	8.5	4	8.5	20	2.5	8.1	0.4	4	5	4	15.76	0	80.76
Me	9.6	5	10	19	9	9.1	1	4	5	4	14.85	1	91.55
Jing	9.4	5	8.7	20	10	9	1	4	5	4	15.83	0.5	92.43
Hua	8.8	5	8.5	17	9.5	8.6	1	4	5	4	18.51	1	90.91
Ming	9.6	5	8.5	16	7	9.8	1	4	4	4	14.15	0	83.05
Yan	8.6	5	8.5	16	9.5	9.8	1	4	5	4	15.79	0.5	87.69
Wen	8.8	4	9.3	13	9.5	9.5	1	4	5	4	17.25	0	85.35
Xiang	0	5	9.3	17	10	0	1	0	5	4	13.66	2	66.96
Quan	9.4	5	9.6	17	10	9.5	1	4	5	4	17.79	0	92.29

Note: I=instruction; M.A.P.=meeting attendance and paperwork; L.P.=lesson planning; P.W.=paperwork; S.W.=student work; S.B.T.R.=school-based teaching research, a collective learning event; W.L.=workload.

Translated by the author.

Only teachers interviewed or observed were included.

[a] The formula is , in which, y is the teacher's final points, 0.5 is the weight, x is the average grade the teacher's class (classes) achieved in the final test, is the district average score of the grade level, and 15 is the base points.

[b] One extra point is awarded for students' exceptional performance in contests.

Teaching norms

Lesson planning

School administrators relied on lesson plans to judge whether teachers had put their heart into lesson preparation. Principal Huang considered it an important indicator of teachers' dedication to and seriousness about teaching. T.D. Zhi also noted that writing detailed lesson plans was more useful for teachers with fewer than 3 years of teaching experience. More experienced teachers did not necessarily write a lengthy lesson plan containing timed, detailed procedures of instruction; instead, they should place central attention on how to teach those major mathematics concepts.

Term teaching plans, unit plans, and lesson plans were three forms of lesson planning. At the beginning of each semester, mathematics teachers would attend the school-wide meeting led by V.P. Yang or T.D. Zhi to re-learn the curriculum standards. Subsequently, together with individual grade-level team leaders, they would go over the whole textbook in order to have a holistic grasp of the knowledge system and the structure of concepts. In so doing, each grade level would compose one common semester plan that should identify clearly the difficult and key points in the text and make a conscientious arrangement of teaching progress over the semester (see Table 8.2).

It is worth noting that 58 periods were used to teach new content according to Teacher Me's spring semester plan, while more than half of lesson time (68 out of 126 periods) was allocated to in-class student work or reviewing that work aimed at consolidating students' learning. It suggested that having student work and reviewing were more important teaching and learning activities.

A unit plan was meant to enable teachers to develop an overall understanding of the whole unit, the objectives to achieve, key and difficult points, time needed for each content area, and the like. Individual lesson plans were the most emphasized type of planning. Pioneer School recommended that teachers prepare lessons at least three days, preferably one week, ahead of instruction. For teachers younger than 35 years of age, they were expected to design detailed lesson plans; teachers between 36 and 45 should have at least half of their lesson plans written in detail; and those over 45 were allowed to write brief lesson plans. It was also specified that teachers of Grades 1–3 should prepare at least 64 periods of lessons, and teachers of Grades 4–6 should prepare 80 periods. Furthermore, in order to promote reflective teaching, teachers were required to engage in active post-instruction reflections, and they should at least have six reflections. Each reflection should contain at least 300 Chinese characters.

As recommended, lesson planning should go through the process of individual, collective, and individual planning. At first, teachers were advised to study carefully various teachers' guides and professional journals, online information, and so on, garner information, and accumulate a solid knowledge base. Following individual planning, teachers at the same grade level should hold one collective lesson planning activity every week. During that event, one person should act as the *lead speaker*, who was expected to first

Table 8.2 Teacher Me's spring semester plan (Grade 2) at Pioneer School

Week	*Date*	*Key content*	*Instruction*[a]
1	March 02–06	Problem-solving: Two step application problems.	4
2	March 09–13	Knowing division.	5
3	March 16–27	Using 2~6 multiplication facts to get quotient.	7
4	March30–April 03	Unit review and assessment.	1
5	April 06–10	Shapes and transformation.	4
6	April 13–17	Using 7~9 multiplication facts to get quotient.	3
7	April 20–24	Problem-solving: Simple division problems.	4
8	April 27–May 1	Unit review and assessment.	2
9	May 04–08	Knowing numbers up to 10,000 (1).	3
10	May 11–15	Knowing numbers up to 10,000 (2).	3
11	May 18–22	Unit review and assessment.	2
12	May 25–29	Gram, kilogram, and measurement.	2
13	June 01–05	+/– of numbers within 10,000.	4
14	June 08–12	+/– of numbers within 10,000.	3
15	June 15–19	Statistics.	3
16	June 22–26	Identifying patterns.	4
17	June 29–July 03	General reviewing.	4
18	July 06–10	General reviewing and final exam.	

Note: Grade 2 had 7 periods of mathematics lessons per week, in total, 126 lesson periods per semester.

[a] Estimated time needed for actual instruction in new content, in total, 58 periods per semester.

explicate his or her understanding of the text's organization of the content area to be discussed, its objectives, and key and difficult points, and lay out a tentative plan of teaching. Other teachers would build upon the discussion. Collective planning purported to "reveal colleagues' wisdom and make it accessible to every teacher" (Pioneer School, 2003b, p.1). Ideally, by capitalizing on the common stock of collegial knowledge, individual teachers could reflect on and perfect their own lesson plans.

A lesson plan should have the preparation date, teaching time, teaching objectives, key and difficult points, instructional aids and manipulatives for students, type of lesson (new lesson, lesson for exercises, review lesson, or experiment lesson, and so on), steps of instruction, prearranged layout of the blackboard, exercises or homework, and post-lesson reflection. All in all, teachers' lesson plans should have *six emphases*:

> [E]mphasizing how to crack key points and difficult points in teaching, emphasizing the effectiveness of learning activities, emphasizing the expectations on students' learning process, emphasizing the design of student work, emphasizing reflections on teaching behaviors, and lastly emphasizing the modification and perfection of expectations.
>
> (Pioneer School, 2003b, p.1)

Pioneer School stressed it was particularly important that lesson plans "be practical rather than formalistic, and be conducive to improving the effectiveness and efficiency of instruction" (Pioneer School, 2003b, p.2). The school evaluated teachers' lesson plans based on eight criteria and a 90-point scale:

> A (12 points): Carefully study the textbook, set accurate learning objectives, and reflect three-dimensional teaching goals, deducting 1 point if objectives were not clear.
>
> B (14 points): Read extensively, grasp the content of the text, and include relevant learning resources to enrich teaching, subtracting 2points if in-depth understanding of the key concepts was not shown.
>
> C (8 points): Design and prepare instructional aid and coursework, taking away 1 point if I.T. was not used.
>
> D (10 points): Carefully design the teaching process which attends to knowledge, skills, and students' affective needs, minus 2–4 points if the process was overly simple.
>
> E (14 points): Use strategies that can stimulate students' enthusiasm and create an equal and free learning environment, taking away 2 points if proper strategies were not evident.
>
> F (8 points): Be conscious of the conflict between the expectation and generation of learning, and remain flexible to adjust instruction according to the real-time unfolding of the lesson, deduct 0.5 point if preparation was not evident.
>
> G (10 points): Demonstrate one's uniqueness and characteristics, and be innovative, minus 1 point per lack of consideration.
>
> H (14 points): Engage in post-lesson reflections or notes which should reflect the reality of the lesson and draw thoughtful conclusions, taking away 0.5–1 point for too superficial reflection.
>
> (Pioneer School, 2003b, p.2)

In order to keep a close eye on teachers' lesson planning, Pioneer School's T.R. Office administered regular evaluation of lesson plans. Approaching the end of each semester, the school also collected all teachers' lesson plans and showcased them in front of all school people. Each lesson plan was also evaluated by the Principal, the V.P., and the T.D. For instance, lesson plans from all 21 mathematics teachers were displayed. Throughout the semester, the T.R. Office selected teachers' lesson plans at least twice without advanced notice. In the spring semester, the school evaluated teachers' lesson plans five times, on February 28, March 24, April 23, May 26, and July 15, respectively. An illustration of the inspection outcome is given in Table 8.3.

It was noted that Teacher Xiang did not write lesson plans, Teacher Me needed to better address how to use I.T. in teaching, and Teacher Wen did not employ diverse approaches to instruction.

Table 8.3 Pioneer School's lesson plan evaluation

Teacher	*Criteria*								*Total (out of 90)*
	A	*B*	*C*	*D*	*E*	*F*	*G*	*H*	
Me	–1		–2				–1		86
Xiang	–12	–14	–8	–10	–14	–8	–10	–14	0
Quan	–1						–1		88
Wen	–1				–2		–1		86

In fact, teachers at Pioneer School had mixed feelings about composing lesson plans. Teacher Xiang did not write a single lesson plan in the school year, for she thought it as a waste of time to write lesson plans. With more than 20 years of teaching experience, she preferred to prepare lessons mentally: Reading the teacher's reference book, going online, and internalizing the text. But, the next year, Teacher Xiang claimed, she would opt to write plans too, since it was "a loss of face" (Teacher Xiang, Interview # 2) to get zero in the evaluation.

Differing from Teacher Xiang, Teacher Me and Teacher Quan recognized the value of writing down thoughts on paper when preparing lessons, since that action helped them to organize concepts and to streamline teaching strategies in their minds. At the very least, Teacher Me stated, it helped her to memorize the knowledge better.

Instructing

Pioneer School had general and particular recommendations for teachers. The recommendations pertained to affective, attitudinal, and behavioral aspects of teaching. Mastering fundamental teaching skills (*jiao xue ji ben gong*; *jiao xue*=teaching, *ji ben*=fundamental, *gong*=skill) and having an accurate grasp of subject matter were also stressed.

Teachers should remain positive and cheerful. In terms of their attitude, teachers should pay attention to the ideas of the N.M.C: Promoting self-initiated, cooperative, and inquiry-based learning, and guiding students to experience the process of knowledge generation, development and formation. Behaviorally, they should arrive at the classroom 2 minutes before the bell rings, and they should not leave the classroom midway through a lesson, instruct while sitting in a chair, teach overtime, or answer a cell phones, or send text messages in class. Nor should teachers leave the class unattended or swap classes without the T.R. Office's permission. Every school day, three teachers on duty would walk around the building to check if teachers were teaching according to the official schedule, or if they were absent, or look for any other issues that might be observed. The observation results were noted on the inspection chart and submitted to the T.R. Office. For instance, on May 08, eight teachers were found swapping classes without permission and one teacher missed his scheduled class.

Teachers were required to master two fundamental skills: Speaking concisely, accurately and loudly in instruction, and writing neatly, formally and systematically on the blackboard. Ambiguity and inaccuracy in the delivery of content knowledge were not allowed. Terms such as "maybe," "perhaps," and "we guess," which reflected a teacher's shallow and uncertain understanding of the subject, must be avoided. Pedagogically, teachers were expected to integrate organically reviewing old knowledge (*fu xi*; *fu*=again, *xi*=practice), teaching the new content (*xin shou*; *xin*=new, *shou*=instruct, teach), and doing exercises to practice (*lian xi*; *lian*=practice, *xi*=practice).

In 2003, the school implemented an updated teaching evaluation system that accommodated the major ideas of the new mathematics curriculum. The system shown in Table 8.4 was based on a 100-point scale; it guided observers to attend to evidence of both teaching and learning practices that reflected the lesson's alignment with the reform.

The evaluation plan attended to a teacher's instructional performance from the perspective of both teaching and learning. Teachers were channeled to include important reform ideas in their lesson planning and instruction, and an emphasis was placed on using I.T. In the meantime, students' responses to the teacher's instruction were equally taken into account.

The school primarily used the plan to evaluate teachers on formal occasions, such as school-wide teaching competitions, instead of everyday instruction. An example was the Competition in Classroom Instruction of Youth Teachers that took place from October 7–17. Nine mathematics teachers took part in the contest and each one was evaluated by three veteran teachers. Teacher Quan received the highest score (286), followed by Teacher Ming (282), and Teacher Jun (274). At the conclusion of the event, the T.R. Office published its review on the nine teachers' lessons. Both the strengths and the weaknesses were noted. In particular, it pinpointed that teachers had done well in creating engaging situations to open their lessons, relating the textbook knowledge to students' lives, allowing students to participate in the process of knowledge formation by using hands-on approaches. Teachers used strategies such as sorting or playing with the objects (*bai yi bai*; *bai*=place, put, *yi*=one, *bai*=place, put), drawing (*hua yi hua*; *hua*=draw, *yi*=one, *hua*=draw), touching (*mo yi mo*; *mo*=touch, *yi*=one, *mo*=touch), observing (*kan yi kan*; *kan*=look, *yi*=one, *kan*=look), and effectively integrated the multimedia in instruction. Half of the document also critically discussed the deficiencies in those teachers' teaching. One of the issues was that teachers could not fully empower students to identify patterns and synthesize concepts on their own, but "always attempted to feed students and answer questions for them" (Pioneer School, 2008, p.2). It also noted that students could not work meaningfully in small groups, which was because teachers had failed to carefully design the questions, timing, task allocation, and the whole process beforehand. Thus, "the input of those academically outstanding students in the groups overshadowed that of the rest, and students who were struggling academically were only a decoration" (Pioneer School, 2008, p.3). The outcomes of the contests were not counted in the year-end teaching evaluation.

Table 8.4 The evaluation standards for elementary mathematics instruction at Pioneer School

Object	*Evaluation items*	*Evaluation*		*Note*		
				O	*G*	*S*
Teacher (50 pts)	Objectives of teaching	Develops students' basic knowledge, skills, and promote positive affection for, attitudes to and beliefs in math.				
	Contents	Bases materials on actual learning and teaching needs, choose and organize content creatively.				
	Process of instruction	Open scenario	Conducive to the unfolding of learning activities and students' cognitive development.			
		Student participation	Students are confident, active, have sufficient time & space to engage in cooperative learning & problem-solving.			
		I.T.	Proper selection and use of technologies.			
		Teacher–student relationship	The teacher acts as organizer, guide, cooperator, and collaborative investigator in students' mathematics learning; equal, harmonious, pleasant, and democratic, the teacher is encouraging and respects every student.			
Student (50 pts)	Affective learning	Is enthusiastic about learning mathematics and curious about things related to mathematics; confident in mastering mathematics and having strong aspirations to investigate and solve mathematics problems.				
	Learning styles	Able to learn independently; haves a good habit of listening attentively, willing to actively collaborate with peers in group learning activities, and responds actively to peers' questions or answers; able to learn mathematics via hands-on activities and from the real life.				
	Learning capacity	Actively finds problems, asks questions, solves problems, articulates the process, and reaches conclusive learning results; capable of accurately assessing peers' learning.				
	Creativity	Sees connection between mathematics and everyday life; thinks freely and dares to express different views.				
Extra: Unique features		Any special/creative features noticed.				
Comments						

Note: O=outstanding, G=good, S=so-so.

Table translated by the author.

Hence, the most common approach to evaluating everyday instruction was that school administrators would directly go into classrooms and observe lessons without notifying teachers in advance, at least twice per semester. Each administrator usually observed more than 50 periods of lessons throughout the whole semester. Lesson observation was mostly used as a preventive measure to keep teachers from cutting corners and remaining accountable to students. Principal Huang considered that direct lesson observation was an important source of information to inform the administrators of whether one teacher performed to a high standard, and whether interventions should be taken, since lesson observations could easily reveal if a teacher's teaching was at issue and on in what respects:

> Suppose two teachers are teaching the same lesson. One teacher teaches 5–10 minutes, then asks students to work on a few problems on their own for 5 minutes, then comes back to address students' questions, and then teaches more, that is, doing exercises along with instruction (*bian jiang bian lian*; *bian*=at the same time, *jiang*=instruct, *lian*=practice), and combining instruction with practice (*jiang lian jie he*; *jiang*=instruct, *lian*=practice, *jie he*=integrate). The other teacher teaches 15 minutes and has students do exercises for the rest of the lesson. Apparently, the second teacher is slothful and irresponsible. Why? Every experienced teacher knows the first one's approach is much more physically and mentally demanding, but more effective and better for student learning.
>
> (Principal Huang, Interview #2)

Then, Principal Huang would approach the less diligent teacher and ask him or her to improve.

Whether teachers engaged students in active learning and thought-provoking interactions was another major concern that administrators attempted to address. On December 19, 2006, V.P. Yang observed Grade 1 Class 2 in the morning, and noted:

The teacher was weak regarding organizing students. The state of student learning was not satisfactory Many students were not focused on learning. The teacher should improve her teaching skills and try to engage students by kindling their interest in learning.

Any misrepresentation of mathematics knowledge of teachers could not go unremarked in lesson observation. The episode that follows illustrated this point:

> T.D. Zhi observed Teacher Ming's lesson on perimeter and area. She took notes on the major interactions and activities that took place between the teacher and students. During instruction, Teacher Ming said to students, "The perimeter and the area of a square, whose side is 4 cm, are equal." T.D. Zhi put that whole sentence down and highlighted it, noting, "wrong." At the post-lesson meeting, T.D. Zhi pointed out

> that knowledge error. "Saying the perimeter and the area are equal is inaccurate. They are two totally different units [of measurement]. Area refers to the measurement of a surface, while perimeter is the unit of length. These two different concepts cannot be compared at all. The area of a 4-cm square is 16 *square centimeters* [emphasis in original], while its perimeter is 16 *centimeters* [emphasis in original].
>
> (T.D. Zhi, Interview #2)

In such cases, school administrators immediately intervened and discussed the issues with the teachers concerned.

In essence, the evaluation of instruction, as T.D. Zhi stressed, was not to judge teachers, "whether one is good or bad," but "provide teachers with necessary guidance from people who are much more experienced, and have had the opportunity to thoroughly study the content and teaching together, to the end of mutually improving teaching quality" (T.D. Zhi, Interview #2). Perhaps owing to this philosophy, most mathematics teachers earned full marks on *instructing* in the evaluation (see Table 8.1).

Student work

Pioneer School looked at student work and exercises as the core component of the whole teaching and learning process. T.D. Zhi considered it the most important task for teachers to design appropriate student work and review turned in assignments. Teachers Wen and Me underscored that teachers could not obtain a clear picture of students' learning state without reading and analyzing student work. Without the feedback from student work, instruction would become pointless. Teachers also relied on student work to decide which students should be tutored.

To alleviate students' heavy workload and maximize the effectiveness of student work, the school asked teachers not to assign written homework for Grades 1–2, and homework for Grades 3–6 should not exceed 40 minutes on weekdays and not over an hour on weekends. In line with the tradition, students should do at least 20 problems per day for computational fluency; in response to the N.M.C., students should design at least four handwritten math posters per semester. The school was strongly against the ocean of problems tactic, that is, employing excessively mechanical and repetitive drill.

Written student work should conform to certain formats. Students were required to put down first the number and then the date of the assignment and begin in a new column. It also stated that teachers must mark student work promptly, conscientiously, and immaculately, using ✓ for correct answers and ○ for wrong answers, and it was highly recommended that teachers mark student work in the presence of the students and the students correct their work on the spot (*mian pi*; *mian*=face, *pi*=mark; *mian gai*; *mian*=face, *gai*=correct). The minimum number of marking instances was also stipulated: Grades 1–2 teachers should correct student work at least 40 times per semester, and

Grades 3–6 teachers at least 50 times per semester. After students corrected the errors, teachers must re-mark the work.

The school evaluated how teachers made use of student work to ensure student learning. Similar to the evaluation of teachers' everyday instruction, Principal Huang stated:

> The purpose of inspecting student work is to tell if teachers are working seriously and diligently. As a veteran teacher, just by browsing through a few pages of students' workbooks, you will have a clear idea as to which teacher is doing the job diligently and which is slack. Did the teacher mark students' work with care, or just rush to get the job done quickly? Did his or her students correct all mistakes? Did the teacher omit any errors that students make? Besides, whoever assigns students a heap of problems, for example, asking students to do problems 1, 2, 3, 4...from the top to the bottom [of the page], is definitely not a good teacher – because he or she is lazy and has no intention to distinguish the types of problems and the amount of student work.
>
> (Principal Huang, Interview #2)

To keep teachers on the right track, Pioneer School selectively examined students' workbooks every month. The T.R. Office randomly picked two or three students' notebooks from each class and evaluated them in accordance with the rubric displayed in Table 8.5.

The T.R. Office inspected how three Grade 4 teachers approached student work on June 7. Six students were sampled from Teacher Yu's two classes, six

Table 8.5 Review rubric for student work at Pioneer School

Category	*Quality standards*	*Measure*	*Score*
Quantity (5 pts)	Meeting grade-level requirements on number of marking opportunities.	Selective examination	–1/violation
Form (5 pts)	Having multiple forms: Self-initiated work, practice-oriented, hands-on work, inquiry.	Selective examination	–1/violation
Standards (5 pts)	(1) Neat handwriting, right format, clear organization. (2) Errors corrected timely. (3) Workbook protected well.	Selective examination	–0.5/student –0.5/student –0.5/student
Marking (5 pts	(1) Marking carefully, timely, with feedback. (2) Re-marking students' corrections. (3) Leaving encouraging comments.	Selective examination	–0.5/student –0.5/student –0.5/student

Note: Translated by the author.

Each point was worth 0.5 points in the final evaluation.

from Teacher Lan's classes, and three from Teacher Quan's class. Over the semester, Teacher Yu's students were asked to complete written work 25–49 times, Teacher Lan's students 53–86 times, and Teacher Quan's students 73–76 times. It was noted that compared to the other two teachers who got full marks, Teacher Yu treated student work carelessly: Insufficient number of assignments (minus 1 point), no dates to show when marking was completed (minus 1.5 points), and illegible comments (minus 0.5 points).

In the spring semester, the school inspected student work seven times, on March 23, April 2, May 11, May 14, May 26, June 3, and June 9, respectively. Taking all seven inspections together, Teacher Me obtained a perfect score owing to her scrupulous treatment of student work. Even though the school administration was satisfied with the majority of teachers, Teacher Ming, a Grade 3 teacher, was an exception. V.P. Yang and T.D. Zhi found that the teacher was not treating students' work very carefully and some low performing students missed a number of assignments. Under such circumstances, the teacher was supposed to provide extra tutoring to those students and to make sure that they turned in schoolwork.

Overall, the school was satisfied with the amount of work that teachers assigned to students. Take Grade 6 classes as an example. From early March to the middle of July, Student Hao in Teacher Quan's class had finished all of the problems in the 90-page textbook, two exercise workbooks (one with 68 pages and the other 90 pages, each page containing roughly ten problems), and 42 timed test practice papers (each paper having approximately 50 problems, allowing 90 minutes maximum to finish). Each practice test was ranked class wide – his ranking ranging from 42nd place (82 points) to 19th place (98 points), and to 1st place (the perfect score). Figure 8.1 illustrates the total amount of student work Student Hao had finished in that semester.

Student Hao completed more than 4,000 problems during the spring semester.

Grades 3 and 6 were the main grade levels that the District T.R. Center monitored in the district-wide Uniform Examination, and their average test grades achieved in the Examination were ranked district-wide. Teachers of these two grade levels tended to give students more test practice papers than others. For example, Teacher Me reported that her Grade 2 students had had fewer than 25 sets of test practice papers in the previous spring semester.

Administrators as well as teachers of mathematics at Pioneer School were particularly proud that their students had much lighter school workload than peers in other local schools. V.P. Yang and T.D. Zhi both acknowledged that Pioneer School was highly regarded for its attention to developing well-rounded children and to alleviating students' academic burden, which was considered one of the school's special features catering to parents. Teacher Me spoke of the difference at length:

> We have tutored students from other schools, like Merits School. They may redo the same test practice paper three times. One is given when the

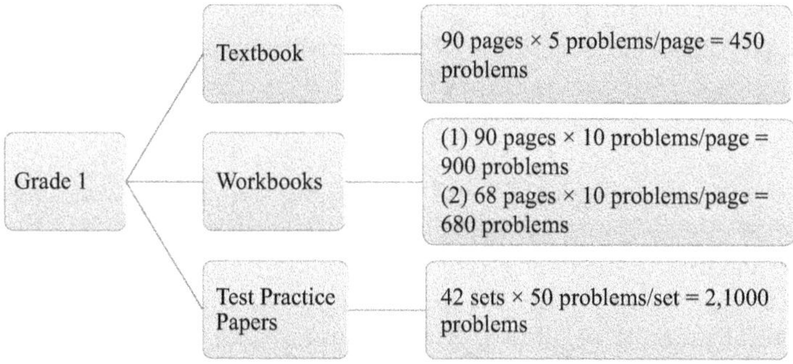

Figure 8.1 Total amount of one student's work at Pioneer School from March to July.

> unit is finished and the other two will be used for final review. Or, they buy three or four volumes of papers and drill repeatedly. Being trained this way, students can surely answer faster and score higher when taking the test. But their thinking is restricted too and they can do nothing beyond the paper. No potential and no characteristics! Why do middle schools like to enroll graduates from Pioneer School? Because our students' potential is not drained. The special feature our school highlights is alleviating academic load. Why do many students transfer into our school every semester? We ask them for what reasons – the load of student work [in other schools] is too heavy!
>
> (Teacher Me, Interview #3)

Too much drill might dummy students because it can "leave no room for students to think, reflect, synthesize, and induce" (Teacher Hua, Interview #3).

Indeed, several Grade 6 students at Pioneer School confirmed that their mathematics workload was not heavy. But, two issues are worth mentioning here. First, teachers were observed not meeting the requirement of having multiple forms of student work. The majority of them still relied on commercial exercise workbooks and test papers full of drill problems. Teacher Hua's Grade 3 students were facing the Uniform Examination in the summer. She had to prepare them for the Examination that only tested knowledge with no concern for real world applications and skills. She explicated:

> Students have to do well in learning pure knowledge to get good grades. The employment pressure is so huge. If they fail tests, their whole life will be finished...It is distorted – the only standard of evaluation is test grades, by one single test paper. Singular assessment, pure test grades, then [students] have to learn pure knowledge, acquiring no skills. Pure knowledge, pure test problems. Because in the test students need to obtain as high test grades as they can, [we] have to be highly efficient in

> normal teaching. One of the highly efficient instructional methods is to ask [absentminded] students to stand up and listen. Without doing test practice papers, [students] don't know which problems they cannot solve and which knowledge areas they haven't mastered.
>
> (Teacher Hua, Interview #3)

Second, even though students' workload at Pioneer School was relatively lower than at other schools, teachers did not necessarily ease their own burden. Most mathematics teachers at Pioneer School taught two classes, about 120 students. Even if it only took 1–2 minutes to grade one student's work per day, the total amount of time was 240 minutes or 4 hours. As observed, Teacher Yan, a Grade 3 teacher, pointed to herself and another colleague, saying "One cow! Two cows! Grading students' homework the whole morning even without raising our heads!" They had been correcting students' work for about 2.5 hours (from 9:00 am to 11:30 am). As most teachers stated, much of their time was spent on this task. It called for teachers' dedication. "This is how we really work day-to-day," Teacher Wen commented, "The key is [you need to] have a sense of responsibility, preparing lessons, marking students' work, and tutoring low achievers...persist in doing these through your lifetime and have patience" (Teacher Wen, Group Conversation).

Quality of teaching

Teachers' quality of teaching was measured on the basis of outcomes of assessment of learning and the achievement of students in contests within or outside the school. To monitor everyday teaching and learning outcomes, teachers were asked to conduct regular unit assessment tests. It was recommended that unit assessment should use different approaches, such as hands-on operations, take-home exams, oral tests, experiments, and so on. After the tests, they should "analyze the results carefully and fully; be aware of the strengths, shortcomings and blind spots of teaching and learning' and come up with effective remedial measures" (Pioneer School, 2003b, p.4). In the end, teachers should turn in the Quality Analysis of Unit Assessment to the school T.R. Office. Neither ranking students according to test grades nor publicizing the test grades was allowed. As for the category of student contests, teachers could get 0.5– 2 points depending on the level of the contest.

The school assigned 4 points to unit tests, 20 points maximum to the final examination, and 1 point to students' achievements in contests (outstanding performance in contests could earn 1 extra point for the teacher). On most occasions, all teachers got 4 points except Teacher Xiang on unit tests. Teacher Xiang gained 2 points because two of her Grade 5 students attained first place in one city-level mathematics contest.

The outcomes of the final examination played the dominant role in evaluating teachers' quality of teaching. Every teacher started with 15 base points and each single point higher or lower than the district average test grade of

Table 8.6 Average test grades of Grade 3 in the Uniform Examination at Pioneer School

Class	*Teacher*	*No. of students*	*Average*	*District average.*	*Difference*
1	Hua	60	92.51	85.50	7.01
2	Ming	57	84.46	85.50	–1.04
3	Ming	58	83.14	85.50	–2.36
4	Yan	59	88.14	85.50	2.64
5	Yan	59	86.02	85.50	0.52
6	Wen	59	89.98	85.50	4.48

the teacher's class average would win or lose an extra half point. In July, all teachers obtained full points in unit tests.

As for the final examination, teachers performed differently. In the school, Teacher Hua and Teacher Quan both far exceeded the district averages (Hua: Average =92.51, district average=85.50; Quan: Average=90.88, district average=85.30). Grade 3 reflects these results (see Table 8.6). Both of Teacher Ming's classes did not reach the district average. Table 8.6 displays these results.

Teacher Ming got the lowest evaluation score among the four teachers, noting that her class average was almost ten points lower than the highest in the grade level. V.P. Yang was unsurprised, since some parents had been complaining about her sloppiness in teaching. Teacher Ming paid a real price for it. Even though she appeared to work hard and turned in well written lesson plans and other required documents, Teacher Ming was removed from teaching mathematics the next fall.

The school did not push teachers for test grades, but it seemed that most teachers still paid a lot of mind to test grades and created peer pressure to some degree. Teacher Wen stated frankly that teachers compared themselves with one another on the final teaching outcome, that is, test grades in final examinations, which reflected not only one's teaching abilities but the degree of one's dedication and diligence. Those whose classes performed poorly in tests generally felt shamed and were afraid of being looked down upon by colleagues. Probably that was true. Teacher Wen and Teacher Me mentioned one of their former male colleagues as an instance. The teacher had been capable of giving superb and lively reform-minded demonstration lessons in front of the public, but his students had always done poorly in final examinations owing to his laziness and inattention to everyday teaching, and he cared little. To this day, he still stood out as someone good at "playing a show".

Test-induced rivalry among colleagues in the school was not apparent. Only one teacher (who insisted on not being identified) disclosed her conflicts with two colleagues. She told us how she was defeated 18 years ago because of collegial rivalry:

> It was my second year of teaching. I was young and taught Grade 1 mathematics. Two days before the end of the semester, with a dozen mathematics

> teachers I gathered in one huge office and graded students' papers. When the average score was calculated, my class got over 89 points, 0.1 or 0.2 less than 90. A senior teacher was responsible for summing up all average test grades of the Grade 1 group. She saw my report card and suggested to me, "Why not make it 90? We will look through your students' test grades and see where you graded too strictly." In that moment, I was holding a handful of poker cards and was very relaxed, so I carelessly agreed. That teacher helped go over my students' papers and found another 12 more points here and there. Even if she hadn't added that 0.2 points, my class would have outperformed most other classes. After we left the school, one of my colleagues immediately went to the Principal and reported that we had cheated and changed students' answers. That teacher threatened to change her test grades too. Overnight, the Principal, V.P., and directors examined all of my students' papers. I was required to turn in a piece of self-criticism the next day. From then on, the whole school knew Teacher so and so made a big mistake…I even thought about suicide. These past 20 years, I have had a very depressing teaching life. Such a small pond, someone churns to make it so muddy.
>
> (Teacher Anonymous, Interviews #1 & 2)

The teacher paid a high price. Her passion for teaching was dampened, her self-esteem damaged, and career advancement thwarted. She held a deep distrust in colleagues who might be insincere. She also mentioned the lack of collaboration by another colleague in the office. She and her colleague had taught mathematics in the school for nearly 20 years. They shared the same office, but her colleague seldom discussed teaching matters with her. When leaving the office, her colleague would lock in drawers all teaching notes, lesson plans, teacher reference books, student work and the like, in case she stole information from her.

Another teacher expressed a less visible form of pressure from her peers. Teacher Hua, a Grade 3 mathematics teacher, had the highest test grade out of her grade-level peers. She felt uneasy, since "they look at you differently. They come to say, how nice, you got number 1. It makes me very uncomfortable. I don't want to be number 1. To be average, not too good or too bad, that is best for me" (Teacher Hua, Talk in Hallway). Despite that nervous feeling, Teacher Hua assured that she and colleagues did not have any problems collaborating.

Parent review

Each semester, the school welcomed parents to attend the School Open Day for Parents. On that day, some parents would choose to be present at the school and observe teachers. Their feedback about teachers' instructional performance was solicited via a survey, concerning three aspects: The evaluation of the teacher's instruction, comments about the state of their child's learning, and what the parent has learned from the activity. In particular, parents were asked to comment if teachers' instruction was in line with the tenets of the curriculum reform. The pedagogy of "stuffing the duck" was strongly disfavored. Parents could also use this opportunity to report on other issues. Nevertheless, since parents were not very knowledgeable of what constituted reform pedagogies, their comments on teachers were largely focused on teachers' diligence and conscientiousness.

Teacher Ming received negative evaluation from parents. V.P. Yang commented with pity that Teacher Ming did not put her whole heart into teaching. While teaching two classes, Teacher Ming also worked as a part-time tour guide, which might have taken her time away from teaching and upset parents. Taking multiple factors together, as a consequence, the school administration decided that Teacher Ming was no longer qualified to teach mathematics and instead was assigned to teach science in the next school year.

Meeting attendance and paperwork

Throughout the semester, teachers were required to attend weekly schoolwide meetings and prepare various written materials to evidence their commitment to the curriculum reform, which would be reviewed by the school T.R. Office. Teachers were required to submit to the school each semester 19 kinds of documents. They included the semester teaching plan, teaching research plan, personal teaching plan, mentor-apprentice plan, peer mutual support plan, personal (research) plan on special topics, personal self-study plan, T.R. group plan on the school-based curriculum, lesson plans for the school-based curriculum, materials prepared for monthly research, teaching research record, self-evaluation on the conformity to teaching norms, speech materials as the central spokesperson (during teaching research activities), quality analysis of unit test, cases of the school-based curriculum, monthly work schedule, and reflections. These written materials were required to be aligned with the reform.

In addition to the above materials, teachers also needed to have two notebooks: One for taking notes during lesson observations and the other for self-study. Teachers who served as group leaders had one extra plan: the T.R. group teaching research plan. Every plan entailed a summary of one or two pages at the end of the semester.

The T.R. Office kept track of what materials teachers had submitted.

For the purpose of evaluation, 5 points were awarded to the subcategory meetings and written materials. In the spring semester, two teachers did not earn full marks. Teacher Hong missed most of the meetings because of illness and therefore got 0.4 points for the subcategory. Teacher Xiang refused to turn in any documents, resulting in 0 points for the subcategory paperwork. Teacher Xiang voiced her strong objection to the school having teachers prepare the excessive volume of paperwork that, in her opinion, was bureaucratic and a waste of time. On this matter, several teachers concurred with Teacher

Xiang and it was observed that some teachers copied from others or used online materials. Nonetheless, Teacher Xiang decided to conform with the school's rule and cobble together something the next year, because getting zero was simply disgraceful.

Behavior

The school's evaluation plan also took teachers' behavior, that is, workload and attendance, into consideration. This category of evaluation was more concerned with teachers' general performance than their reform-aligned instructional change. According to the regulation of the C.E.B., the official full workload for mathematics teachers was 14 periods of lessons per week. In the final evaluation, the school assigned 20 points to the subcategory of workload. Each period of lessons was awarded 1.5 points. Teachers teaching across grades or subject areas would receive 1 or 2 extra points, respectively. For teachers who had a class larger than the official maximum class size of 50 students, every five more students would be worth one extra point. Take Teacher Wen in Table 8.1 as an example During the school year, Teacher Wen taught only one Grade 3 class, a total of eight periods of lessons per week, or six periods fewer than the official workload, so she lost 9 points. However, since she had 59 students in the class, or nine students more than the official class size, two extra points were credited to her, that was, a total of 13 points in the evaluation.

The school T.R. Office kept a detailed record of teachers' daily attendance, which counted the number of days that teachers were absent, on medical leave, or did not sign the attendance sheet in the morning. A total of 10 points were assigned to the subcategory of attendance. For each absence, the teacher would lose 0.25 points, and for every two occurrences of not signing in, another 0.25 points would be deducted. For instance, during the school year, Teacher Me missed two days, did not sign in five times, and got 9 points for attendance in the final evaluation, while Teacher Hong missed 17 days and was late for 26 times so he earned 2.5 points.

Indeed, this category of evaluation was less about teachers' change in curriculum reform. Yet, one negative consequence was notable, as Teacher Zhou and Teacher Xiang mentioned: Because most teachers had overly large class sizes and stated that they were overloaded, it was very difficult for them to carry out reform-based pedagogy such as using small groups in instruction. While they were compensated for having large classes in the evaluation, it could not effectively boost teachers' enthusiasm or capacity to perform reform-based teaching strategies.

School-based teaching research

S.B.T.R. (*xiao ben jiao yan*; *xiao*=school, *ben*=base, *jiao*=teach, *yan*=research or study) was emphasized by the school administration as an effective professional development approach to prompt teachers to transform their knowledge and instructional practices. S.B.T.R. took place at the school, in the grade-level T.R. group, and at individual levels, taking at least 30 hours, 28 hours, and 32 hours, respectively. every school year.

The school level teaching research was more general and aimed at all subject areas. Usually, teachers gathered the week before the beginning of a new semester and undertook a variety of training activities. For example, on February 25, teachers watched the video about the deeds of Teacher Xiaoawe Ren, one national exemplary teacher in moral education, and Principal Huang made a speech on how to be a successful teacher of high moral caliber; on the following day, February 26, the school held a workshop on how to understand and utilize new curriculum standards in teaching. During the meeting, V.P. Yang explicated on the purpose, goals, and tenets of curriculum standards, and pointed out the significance of aligning teaching with specific standards. Moreover, every Tuesday afternoon, two lesson periods were allocated to schoolwide collective learning. Topics ranged from training teachers to use software applications or technologies such as the interactive whiteboard to enhance teachers' professional ethics.

As for grade-level T.R. group-based teaching research, two lesson periods on Thursday afternoons were used for collective learning activities such as lesson planning, analyzing the textbook, and lesson narration, and so on. At the beginning of the semester, the group leader would plan out weekly discussion topics and particular teachers were assigned as the lead person. Throughout the semester, every teacher was supposed to serve at least twice as the lead person. In addition, the school regularly organized group-based lesson demonstrations, open lessons, and lesson competitions. Another important event was the Teaching Research Month that took place in October or November. During this month, each mathematics teacher in the grade-level T.R. group would prepare one lesson, teach the lesson in front of the other members, participate in post-lesson critique, and refine the lesson. The event often was completed within 2 weeks and ended up with having one model lesson from each grade-level T.R. group.

Teachers were also required to engage regularly in individualized self-study. They should make teaching research plans on special topics at the beginning of the semester. Their self-learning activities ranged from reading professional journals or education magazines and taking notes to writing up reflections on readings, to analyzing typical cases of instruction. In addition, learning from peers was also emphasized by the school. It was specified that mathematics teachers should observe at least 15 lessons given by their colleagues.

Various teachers stated that S.B.T.R. was one of the most productive and effective means for them to learn from peers and keep up with the latest trend of educational reform, especially through such activities as lesson demonstration, lesson observation, and critique. Teachers benefited most from lesson-related S.B.T.R. Yet, most teachers were less favorable to the amount of time spent on paperwork.

The school had four subcategories on teachers' S.B.T.R.: Participation in teaching research activities (15 points), professional learning (5 points), individual learning on special topics (5 points), and paperwork (5 points). Teachers were gauged accordingly and awarded specific grades. The grand total multiplied by 0.3 was used to obtain the final score in the teaching evaluation. For the first category, anyone who missed any schoolwide learning or group-based teaching research activity lose 1 point. Those absent because of school matters were not punished. In order to keep a close eye on teachers' S.B.T.R., all school administrators were required to participate in weekly T.R. group meetings and review teachers' written materials.

To standardize teachers' professional learning, the school asked teachers to make the individualized teaching research plan (2 points). At the least, everyone should read one educational book, write one reading reflection, and turn in the report on the analysis of one typical instructional case. Teachers should take notes while reading professional materials at least 15 times during the semester. The quality of plans and written materials turned in was worth 3 points.

As for individual learning on special topics, teachers needed to select one topic at the start of the semester to enable deeper self-learning via research linked to their everyday teaching practice. Teachers were supposed to choose the topic based on the reality of their teaching and students' characteristics, and learn to search literature, collect data, and write up research results in the end. The initial plan and the final report were each worth 2 and 3 points.

The last category emphasized the completeness of five types of written materials on S.B.T.R. Each teacher had a notebook entitled Record of Teaching Research Activities, which included the individual teaching research plan, the final report, the analysis of the typical teaching case, the written materials they created when they were the lead person, and the research report on special topics. Each was awarded 1 point. At the end of the semester, administrators would review the notebook and assign grades accordingly.

The school placed a greater emphasis on the notebook. If teachers did not turn in the Record of Teaching Research Activities, they would get zero for the category of S.B.T.R. This was the case of Teacher Xiang in the teaching evaluation: She did not have any written records to show her participation in teaching research activities and thus obtained zero in the S.B.T.R. Teacher Hong missed five school-wide and group-based teaching research events, so he lost 5 points. The school also took a half point off his written materials. Thus, his raw score was 24.5 points or 8.1 points after applying weighting.

Summary

The reform-minded teaching evaluation system institutionalized at Pioneer School played a critical role in effecting curriculum reform. The system served both managerial and developmental purposes. On the one hand, it was implemented as a time-consuming administrative tool to ensure that teachers would perform in line with the reform ideas and engage in quality teaching. On the other hand, most teachers conscientiously adapted their teaching behaviors. Several shortcomings of the evaluation system were mentioned by teachers.

9 Teachers' experience of the curriculum reform at Pioneer School

Teachers talked about change

Changed views of mathematics and mathematics education

Even though educational change might not have been as progressive as reformers had envisioned, mathematics teaching and learning at Pioneer School had demonstrated considerable differences from before. After nearly 10 years' participation in the reform, teachers had developed markedly different conceptions of what constituted mathematics and what mathematics meant to teachers and students. Before the reform, teachers looked at mathematics narrowly as something related to numbers. As Teacher Me explicated:

> Before, we said what concerned numbers is mathematics. We thought we could solve all mathematical problems by working with numbers. the N.M.C. changed that notion. The definition of mathematics is deepened. Mathematics does not only concern numbers. Many problems that might not involve numbers are also mathematical. Nowadays, we have problems like telling time and money, identifying symmetry, rotating, and statistics. Its scope is much wider.
>
> (Teacher Me, Group Conversation)

Owing to the reform, the reformed curriculum encompassed such contents as geometry, statistics, and probability. Teachers gained greater exposure to knowledge that was not in the realm of traditional elementary mathematics. It needs to be mentioned here that student academic achievement at Merits School, measured by test grades, was much better than that at Pioneer School, as Table A2 shows.

Along with the expansion of the scope of mathematics, teachers saw teaching mathematics differently, too. Nowadays, teaching mathematics meant something more than just calculating numbers. The statement of three-dimensional objectives in mathematics education was frequently mentioned by teachers. As Teacher Me explained, those objectives include the objectives of knowledge and skills, the objectives of [learning] process and methods, and the objective of [students'] emotion, attitude and values. She went on to state:

> The third objective is at least we should have students participate in learning happily, not miserably. So, you need to create an atmosphere and a scenario that grabs students' interest. The second objective is students should learn the knowledge. It is not that we should indoctrinate them with the knowledge; rather, we need to have students learn the strategy of learning. They can use the strategy to solve actual problems in their real life. This is the first objective of knowledge and skills: If we learn the knowledge of identifying patterns, we will be able to use it in work or daily living. For instance, if we are going to decorate the playground with 100 color flags in the pattern of two red, one blue, two red, one blue... and so on. How should we place the purchase order? Again, if we want to pave the patio at home in the order of one red tile, one blue, one green, and one red, for the first line, and then place the last tile in the first for the second line, what will the third line be like? He will know how to apply his mathematics to this work.
>
> (Teacher Me, Interview #3)

Becoming more aware of the emotional needs of students, Teacher Me and her colleague apparently placed a stronger emphasis on promoting the affective dimension of learning. They managed to create engaging scenarios to make students enthusiastic to participate. Over the years, they had been able to develop a repertoire of tactics to stimulate students' enthusiasm for learning. For lower grade levels, Teacher Me often started with telling a story, assigning an imaginative task, or presenting a picture, which engaged students in learning to process information and conduct mathematical analyses. For more advanced grades, teachers would provide students with more intellectually challenging and sophisticated problems that would reveal "the charm of mathematics" (Teacher Me, Group Conversation) to stir up students' interest in learning.

What's more, the utility of mathematics in real life was stressed in teaching. Mathematics was learned for a purpose – being utilized in life. Teacher Xiang stated:

> We think the way that we learned math in the past was futile...too rigid. Math and life were taken as two totally different things. [That way], it was only when we grew up we realized that this piece of math was used. The math we learn nowadays has really changed. The curriculum reform is successful in this sense. [What we learn and teach] can really be useful and most can be connected to or applied in real life, like shopping, making plans or making investments. Students learn to see math this way from Grade 1.
>
> (Teacher Xiang, Group Conversation)

Teacher Wen, Teacher Hua and Teacher Me expressed similar opinions. Teacher Quan insisted that mathematics should be able to feed back to life:

> Why do we learn math? What is the function of math? We realize that daily living cannot do without math. The purpose of learning math is to serve our life. So, in teaching, we place a great emphasis on strengthening the connection between living and math. At the same time, we play down testing.
>
> (Teacher Quan, Interview #2)

In agreement, Teacher Hua stressed, "If something cannot be used in life, what is the value of learning it?" (Teacher Hua, Interview #2).

Since the implementation of the reform, Teacher Me observed a sharp difference between her contemporary students and those she had taught in previous years. Her contemporary students exhibited an apparently higher degree of creativity and self-assurance. Teacher Me noted that her current students could find an answer or solution to any problem given to them, when former students would most probably have left it blank. On most occasions, her students could reasonably justify their solutions. One student was observed in Teacher Me's Grade 2 class:

> The lesson is about identifying patterns. Now, Teacher Me asks students to work on the problems in the student exercise handbook. Five problems are given – (1) 11, 15, 19, 23, (), …; (2) 3, 6, 12, 24, (), …; (3) 2, 3, 5, 9, 17, ( ), …; (4) 1, 3, 4, 7, 11, (), …; (5) 3, 4, 7, 12, 19, 28, (), … – for students to figure out the number in the parentheses. Students start solving the problems individually. Several minutes later, Student Yan gets stuck on the third problem. She turns to her neighbor, Student Huang, and asks him how to solve the problem. Student Huang thinks to himself for a while and adds 8 between the number 9 and 17. He whispers to Yan, "See, 2 plus 3 is 5, 3 plus 5 is 8, 8 plus 9 is 17. So we add 8."
>
> (Teacher Me, Observation #1)

Even though the student did not solve the problem correctly, Teacher Me recognized his different thinking as well as the courage to present his ideas openly.

Changed approaches to teaching

Alongside changed views of mathematics and its learning and teaching, teachers at Pioneer School appeared to be versed in multiple reformed instructional approaches. Teachers encouraged students to think and reason independently. They were comfortable with using small groups and carrying out hands-on activities. During instruction, they gave students ample opportunities to observe, explore, and interact.

Encouraging independent thinking and reasoning. Teacher Me pinpointed that one important aim of mathematics learning was enabling students to think independently and originally, instead of simply imitating others' ideas. "Why do we have no Nobel Prize winners in China? It is partly because we lack the ability to develop original ideas," she emphasized (Interview #3). In Teacher Me's words, students should "have the ability to think from multiple facets" (Interview #3). In order to cultivate this ability, she encouraged her students to employ multiple strategies to solve problems. She maintained:

> If we conformed to one mode of teaching, students would be less capable to think elastically. So we ask that this student could think from this angle and that student from that angle. Everybody can develop unique ideas. That way, students are not thinking linearly but in a three-dimensional fashion. [A problem is] like a hexahedron that has six faces, students could tackle the problem from any face. Even a simple math problem might have end up with four or five different solutions. In general, we have three levels in teaching. The first level is we are satisfied with knowing how to solve the problem. The next level will see if we can have multiple strategies. Some students might find seven or eight ways. Then, the third level is we need to teach students to choose the best solutions.
>
> (Teacher Me, Interview #3)

When Teacher Quan started teaching, the reform had just been embarked on. In her first year of teaching, she insisted that there was only one right way to do mathematics, and students were forced to conform to her thinking. Since the school initiated the experiment in early 2000, Teacher Quan was actively involved in various reform-minded professional development activities. She gradually liberated her conception of learning and teaching and encouraged students to seek multiple solutions:

> Now, we don't set restrictions beforehand. For example, 9+5. Some students will just count from 1 up to 14, while some might count from 9 and 5 more. That works too. Some students might come up with simpler strategies: Breaking 9 into 4 and 5 and regrouping two 5's, or breaking 5 into 1 and 4 and regrouping 1 and 9. Either strategy will do. Then, we will guide students to choose optimal strategies. Actually, students can gradually find which way is simpler after a while. We think this teaching method is more suited to different levels of students. If we forced them to use one particular strategy like regrouping 9 and 1, some students would probably not be able to learn it.
>
> (Teacher Quan, Interview #1)

Using hands-on activities. One prominent change that Teacher Tao observed was they used more hands-on activities in everyday teaching, within and outside the classroom. Teachers asked students to collect data about their families or community, make cutouts, measure real life objects. He acknowledged that students were given more opportunities to connect mathematics to their lives:

> Before, we asked students how many kilograms one ton is equal to. We didn't ask how heavy it is. Just one answer would be sufficient. Thus, we used to solve problems really quickly: Tell students 1 ton[1] equates to 1,000 kilograms and then practice with more problems. In that case, some students would often make such mistakes as one bag of rice was equal to 1 ton. Nowadays, we have changed completely and placed a greater focus on the process of learning, the process of thinking and reasoning. We ask students to think and inquire most of the time. Instead of asking them to tell you the quantity of one ton, we guide students to develop a tangible feeling of how heavy 1 ton might be. We ask students to bring from home one bag of salt, sugar or flour that weighs 1 kilogram. They feel it and experience tangibly how heavy it is.
>
> (Teacher Tao, Interview #1)

Teacher Me shared similar experiences:

> We bring small rods to students to work with and help them experience the length of two sides of a triangle is greater than the third side. In another case: The area of a triangle. If we simply tell students the formula one half times the base times the height, they will easily forget it or forget to multiply one half. That is convenient and effective and saves time, but they don't experience the process of knowledge generation. If students cut two equivalent triangles and make them one parallelogram, they will understand why the area of a triangle is one half of the area of a parallelogram. Learning by rote memorization is fast, but students cannot further their thinking. If we allow them to explore, practice and draw conclusions throughout the process of learning, students will have a better understanding of the topic.
>
> (Teacher Me, Interview #3)

Via hands-on learning, Teacher Quan believed that students were not taught or given readily the piece of knowledge; instead, they could generate the knowledge on their own.

> We taught knowledge to kids before. Now, we really emphasize the idea of letting kids themselves generate knowledge in the hands-on process. For example, finding the volume of a cylinder. After they learn the volumes of cubes and solid prisms, they will be able to use what they have already known and generate new knowledge. We have one type of manipulative: the cylinder is cut into many equivalent sectors. Approximately, every two pieces will be able to form one small rectangular cube. Altogether, it is one big rectangular cube. Through such transformation, students will

> know to what part of the cylinder the width, length and height of the cube are equivalent. This case pinpoints the principle of giving students tools to fish instead of giving them the fish.
>
> (Teacher Quan, Interview #2)

To get students to generate knowledge themselves, teachers started to assign more life-related tasks as homework. As Teacher Hua stated:

> Students were very interested in tasks relevant to their lives. Take learning meter as an example. Many students don't have a clear understanding of the concept of a meter. Normally, they don't use it much. If we told them in class how long 1 meter was, it would take many lessons for them to really get at it. Then, they would probably still have no tangible sense. If a problem like "the height of a tree is 3 what?", some students would put 3 centimeters. They had no experience. So, in teaching, we will ask them to measure the height of themselves and their parents, the length and width of their living room, and the height of trees, with the assistance of their parents. By doing so, they would develop a better knowledge of a meter. They would realize even a small tree would be taller than 3 meters. This way, students can really tell the difference between 1 centimeter and 1 meter.
>
> (Teacher Hua, Interview #1)

What enabled teachers to change?

In order to effectively implement the reform in the school, teachers informed us that they participated in a variety of structured activities that occurred inside and outside Pioneer School. Those activities consisted of visiting partner schools in more economically developed cities such as Beijing, attending school-level discussions and lesson studies, and carrying out group-based teaching research. In both formal and informal ways, teachers could frequently help one another, and improve their knowledge and instructional skills. These activities played significant roles in enabling teachers to learn reform pedagogies.

Learning outside the City

Learning outside the City was the primary means by which teachers at Pioneer School gained access to reform-related information and knowledge. They were provided with plenty of learning opportunities outside, especially before 2006, when the school had not yet been downgraded. The schools they visited were generally more developed economically and more progressive pedagogically than those in the City. Teachers valued such experiences because they could really see teaching different from their own. As Teacher Quan commented, teachers at Pioneer School were "already the best in the city" and quite familiar with one another's teaching styles; they would not have benefited much simply from observing their own colleagues (Group Interview #1).

Learning outside constituted the important starting point for experiment teachers at Pioneer School to learn and implement the N.M.C. They paid frequent visits to schools in Tianjin in early 2000 and readily taken approaches back to their home school. Taking as an example one of their visits to Tianjin in 2001, Teacher Wen shared her thoughts:

> Their teaching reflected reform essence, which was thought-provoking to us. Their students conducted investigations pretty much on their own. Their teachers did not interfere with students' activities much. We realized that we had to let students try. They [teachers] used creative problems related to real life.
>
> (Teacher Wen, Group Interview #1)

Regarding the same study trip, Teacher Quan concurred that students whom they had observed were encouraged to solve problems in different ways. Teacher Xiang also noted that the classes they visited had been active but not chaotic, and teachers had habitually asked students to fully express their opinions (Group Interview #1).

Taking advantage of their frequent visits to other schools, teachers at Pioneer School had greater exposure to reform pedagogy beyond their familiar schools. In addition, through a series of school-based events, teachers translated those approaches to the school. One of the events was school-level learning.

School-level learning with coaches

Teachers at Pioneer School were also engaged in school-level learning activities. Inviting the City T.R. Office fellows to guide their learning was an essential component of teachers' professional development. Since fellows from the city center were mostly more experienced and reform-minded, they positively fostered teachers to learn new pedagogies. School-level learning with coaches took different forms. It could be collective lesson demonstrations, observations, or simply group discussions. An example was shown as recorded in detail during one school-level learning event (Pioneer School, 2004). Fellow Xu asked experimental teachers to share what change they and their students had experienced:

> Teacher Yu said, "the N.M.C. asks teachers to observe, express, and think from multiple angles. Alleviating the burden of students. Focusing on cultivating students' hands-on abilities. But we lack manipulatives and posters." Then, Teacher Quan stated, "The curriculum best suits students at the middle level and above. The old curriculum dictated teachers to

teach *why* explicitly, and the N.M.C. mainly asks students to explore and explain *why*. For example, the Grade 1 textbook has one problem asking students to figure out if the bridge can afford the weight of a truck with different loads of goods..." Teacher Yao added, "The merits of this curriculum are: (a) Nurturing students' abilities to think out-of-box and (b) less content. The shortcoming is lacking exercises to enhance calculation accuracy." Teacher Xiang followed, "We lack accompanying student workbooks." Teacher Lan responded, "The curriculum is closer to students' lives and more inquiry-based and feasible. Students very much like its games such as 'guessing my number'. Students can explore in activities, which is very beneficial to inspire students' interest in learning." Teacher Wen concluded, "[The curriculum] can cultivate students' abilities in language, thinking, and creativity. It offers them the context of communication."

Fellow Xu said, "We think our teachers are not serious enough...From what you just shared, we can tell that you haven't truly delved into the curriculum or understood it. What are the features of the N.M.C.? Have you embodied them in your own teaching? Here are the three features of the new textbooks: (1) Why does the N.M.C. have more pictures? It embodies the tenet of *knowledge comes from life*. The N.M.C. purports to have students understand that their lives are full of mathematics, thus stimulating them to tackle problems in real life; (2) the N.M.C. grants students the owner status in learning. Thus, students at varied levels can do and achieve differently in mathematics; (3) [The curriculum promotes] inquiry-based learning. It emphasizes that teachers and students are equal, and embodies that teachers are guides and collaborators in learning. As for mathematical fluency, we should strengthen drill in this matter. Basic oral arithmetic should be as good as before.

(Pioneer School, 2004, pp.3–6)

In the above exchange, teachers voiced their views of the N.M.C. and displayed how deep their understanding of the reform ideas was. In turn, the fellow critically commented on those teachers' opinions and did not hesitate to point out the shallowness of their sharing. In the end, he recapped the major ideas of the reform and asked teachers to be mindful of both reform pedagogy and mathematical fluency.

As various teachers expressed, because of the presence of someone more knowledgeable, guided school-level learning activities could particularly goad teachers to think deeper so as to demonstrate the best of their knowledge. It became the opportunity for one another to truly share and appreciate what others contributed in the conversations. However, not all teachers had the privilege to take part in such activities. For regular teachers, group-based teaching research was where their learning mainly took place.

Group-based teaching research

Weekly teaching research. Teachers spoke highly of group-based learning. One important event was *weekly teaching research*. Pioneer School reserved the last two lessons of each Tuesday for teachers of the same group to learn together. At the beginning of each semester, the group leader of each grade group would need to compose the semester plan for weekly teaching research. For her Grade 2 group, Teacher Me designed the plan shown in Table 9.1.

With designated times and key speakers, the table lists the topics that Teacher Me and her Grade 2 colleagues were to address every week. Teacher Hua stated that teachers did not necessarily stick to this plan in weekly teaching research. They could select topics that were more suitable to their actual teaching and learning needs. Like others, Teacher Me carried out this activity conscientiously.

The primary benefit was that this arrangement gave her and her colleagues opportunities to interact in a more systematic fashion. In addition to this weekly activity, they participated in some more structured learning, such as the *teaching research month*.

Teaching research month. A specially structured event at Pioneer School was called the *teaching research month,* which was conducted school-wide once every semester. The event usually started in the 3rd or 4th week of each semester and lasted about 3 weeks. Each teacher in the grade group would design a lesson, narrate the lesson to peers in the group, instruct it in front of others, and reflect on it together as a group.

Table 9.1 Example of weekly research topics and key speakers at Pioneer School

Date	*Activity theme*	*Key speaker*
09.04	Studying the N.M.C. Standards.	Mi
09.11	Reviewing the textbook.	Jing
09.18	Semester lesson planning.	Hong
09.25	Group discussion	Mi
10.9	Special topic: Lesson critique.	Jing
10.16	Special topic: summary.	Hong
10.23	Individualized lesson planning.	Mi
10.30	Group learning.	Jing
11.06	Reflection on teaching.	Hong
11.13	Analyzing lesson plan on understanding lines.	Mi
11.20	Design on axially symmetric shapes.	Jing
11.27	Exploration of the lesson of mathematics activity.	Hong
12.04	Special topic on how to teach students to ask questions.	Mi
12.11	Special topic on effective instruction.	Jing
12.18	Special topic on skilfully designing exercises.	Hong
12.25	Collective learning on recurrent problems.	Mi
01.02	Reflection and interaction.	Jing
01.09	Review and preparation for the final examination.	Hong

Note: Translated from Teacher Me's Record of Teaching and Research Activities.

Table 9.2 Teacher Wen's lesson plan

Subject	*How long is the decoration?*	*Grade*	*2*	*Date of planning*	*03/16*
Type	New content	Teacher	Wen	Date of instruction	03/18
Learning objectives	(1) In relation to specific scenarios, exploring and mastering the methods of calculating the perimeters of rectangles and squares. (2) Being able to calculate the perimeter of a rectangle correctly.				
Focal point	Guiding students to explore the methods of calculating the perimeter of a rectangle.				
Difficult point	Arriving at the formula of a rectangle's perimeter.				
Manipulatives	Plastic decoration, meter tapes, rods.				
Time	40 minutes.				
Instructional design					
Steps	Instructional process. (1) Setting the context: Teacher takes out the decoration, asking students what it is and which holiday they connect it to (New Year's Eve). Now, we are going to decorate the border of the blackboard. How long does the decoration have to be? (i) Writing *How Long is the Decoration?* on the blackboard. (ii) Asking students to ask questions. (What information do we need to know? What is the length and the width?) (iii) Asking four students to collaborate to measure the length and width of the blackboard. (2) Now that we have got the information, the class can start calculating its perimeter (by asking what shape the backboard is). (3) Discussing methods (calculating independently and then discussing as a team). (i) length + width + length + width. (ii) length * 2 + width * 2. (iii) (length + width) * 2. (4) Working on two exercise problems on page 45. (5) Discussing the methods (which one do you like?).(6) Practical application: Giving mathematics problems in real life. E.g. if the sides of the roof of one's room (length=9 meters, width =5 meters) are going to be decorated with wooden blocks, how long will the blocks be? How many blocks (each block is 3 meters)? How much money will be needed (each block costs $28)?				
Reflection	(1) Through exploration, having students experience and own the whole process of calculating the perimeter of a rectangle. (2) Through practical applications, students are aware of mathematics in life.				

Note: Translated by the author from Pioneer School (2003c).

Take one former event as an example. From March 15 to April 1, 2004, Teacher Wen, Teacher Me, and another three teachers in the Grade 2 T.R. group were engaged in the monthly event. First of all, Teacher Wen designed her lesson on the perimeter of a rectangle. It is shown in Table 9.2.

Table 9.3 Narration of Teacher Wen's lesson plan

Topic	*How long is the decoration?*		
Coordinator	Teacher Wen	Date of study	March 17
People	Teacher Me, Teacher Jia, Teacher Lin, Teacher Yu		
Content and process			
(Teacher Wen recounted the main points in her lesson plan as shown in Table 9.2.)			
Comments:			
Teacher Lin: Under the guidance of the teacher, this lesson will allow students to understand the rich mathematics information and wide applications in real life. Through the task of decorating the blackboard, students will explore different methods of computing the perimeters of rectangles. They will be able to experience how mathematics is generated. That is positive. The shortcoming is that the New Year activity might be unrealistic for this time of the year.			
Teacher Me: The exercise problems are skilfully designed, very practical and thought-provoking. It integrates knowledge of perimeters and multiplication. Since the teacher intentionally gives little information about the blackboard, students are prompted to ask questions. The design will be conducive to the cultivation of students' thinking abilities.			

The plan shown in Table 9.2 contains the objectives of the lesson, pinpointed its focal and difficult points, and delineated the procedures of instruction. Teacher Wen divided instruction into three major steps and anticipated possible questions and solutions that students might bring up.

The day after the lesson (March 17, 2004), Teacher Wen read out her lesson plan in front of the other four teachers. She explained her rationale for designing the lesson this way and described the instruction process in detail. Then, her colleagues shared their opinions on the plan, as Table 9.3 shows.

This practice is called *lesson narration* (*shuo ke*; *shuo*=speak, *ke*=lesson). Without referring to her lesson plan, Teacher Wen explained the rationales for her design and recounted her anticipated instruction step-by-step to her colleagues in the group.

On March 18, Teacher Wen gave the lesson, while her colleagues were present and observed her teaching. Each teacher was equipped with the Lesson Observation Notebook and kept a detailed record of her teaching activities. Teacher Me's comments appear in Table 9.4.

Teacher Me kept track of the lesson's process. In the right column, she wrote down comments on the strengths and merits of Teacher Wen's instruction. Those comments were shared with others in the post-lesson discussion.

Immediately after instruction, Teacher Wen evaluated her own teaching. She filled out another form, the Self-assessment of Instruction Form, as shown in Table 9.5.

Teacher Wen rated her own instruction highly. She considered that her lesson fulfilled all five criteria of assessment. In her post-instruction reflection, she stressed the success of the lesson. However, in addition to commenting

Table 9.4 Lesson observation record at Pioneer School

Teacher	*Wen*	*Subject*	*Math*	*Grade*	*Grade 2 Class 7*

Topic Perimeter of a rectangle	Comments
1. Creating the context. (1) Teacher proposes the task of decorating the blackboard for the New Year celebration, and asks students to find out how much plastic decoration is needed. Students think and ask questions about the lengths of the four sides of the board. (2) Students collaborate to measure the length and width of the board (four students). Length=320 cm, width=120 cm.	When measuring the length of the board, four students work together. It helps raise their awareness of collaboration. Meanwhile, students experience the process of learning in person, and improve their hands-on abilities.
2. Explorative learning. (1) Computing the length of decoration: (i) Students work independently, and (ii) students report on their methods (four methods). (2) Experiencing further. Teacher gives one more problem: Length=27 cm, width=15 cm. Students work alone and then report: (i) (27 + 15) * 2, and (ii) 27 * 2 + 15 * 2.	Teacher lets students explore the methods, which embodies the reform spirit. It is good for cultivating students' abilities to self-explore.
3. Identifying the pattern. Through induction, students derive the formula.	Perimeter = (L + W) * 2
4. Actual applications (1) Teacher gives the problem: To decorate the walls of the classroom with wooden blocks, how long is needed? Remind students of the previous case. Students solve the problem: (9 + 6) * 2 = 30 m (2) If each block is 3 meters long, how many blocks are needed? Students' solution: 30 / 3 = 10 pieces. (3) If each piece costs \$28, how much money does it cost? Students' solution: 10 * 28 = \$280.	The life-related exercise engages students in exploring mathematics around them. They realize that what they have learned can be applied in real life.

favorably about the lesson, her colleagues identified several shortcomings of her lesson, as displayed in Table 9.6.

Teacher Jia considered Teacher Wen's instruction as having satisfied all five criteria. On the other hand, it seemed that teachers were not keen to offer critical feedback to their colleagues, as Teacher Jia did. Indeed, if teachers had "talked and did too much for students," students would not have been engaged in authentic inquiry-based learning. Furthermore, the teachers' comments also demonstrated that they were not foreign to reform ideas, language, or instructional practices. In fact, those evaluation criteria of instruction were closely reform-aligned and meant to foster teachers' instructional change.

Table 9.5 Self-assessment of instruction

Teacher	*Wen*	*Topic*	*How long is the decoration?*	
Date	03/18	Grade	2/7	Self-assessment
Criteria: 1. In class, the teacher and students are harmonious emotionally, and there are active interactions. (Yes) 2. Self-initiated, cooperative, inquiry-based learning strategies are reflected in instruction. (Yes) 3. The teacher respects students' individual differences, and encourages all students to fully participate in learning. (Yes) 4. The design has the characteristics of creativity, flexibility, and openness. (Yes) 5. Instruction brings related curricular resources. (Yes) Reform-aligned behaviors are: (1) Encouraged students to solve problems on their own, (2) created the problem context to engage students, (3) enabled students to use new knowledge to solve problems, and (4) created suitable scenarios to inspire students to inquire.				
Post-instruction reflection	I capitalized on the nature of children's curiosity and guided students to explore the length and width of the blackboard. It was good to lay the context for further exploration. I encouraged students to speak and give different solutions.			

Table 9.6 Peer evaluation of instruction

Teacher	*Wen*	*Topic*	*How long is the decoration?*		
Date	03/18	Grade	2/7	Observer	Jia
Criteria: (The same five criteria as those in Table 9.5.) Reform-aligned behaviors are: (1) Had students collaborate in groups and conduct hands-on activities, (2) enriched students' understanding of multiple measuring methods, (3) explored multiple solutions and emphasized the optimal solution(s). Strengths: The teacher respected the ownership of students in learning and asked students to find the optimal solutions. Shortcomings: (1) The teacher talked and did too much for students; (2) the teacher should have asked students to discover other plane shapes in life and think about methods to calculate the perimeters.					

As one of the key teaching research events at Pioneer School, teaching research month effectively "spread new thinking and things" (Teacher Me, Interview #3) with the group and schoolwide. In particular, inexperienced teachers or those new to reform ideas oftentimes were given the most attention during this structured event. As a result, teachers' instructional change in the school had taken place over time.

Obstacles to instructional change

Teachers developed tremendously after many years into the reform. In general, their views of the reform and their own growth were positive. Nevertheless, some mentioned several obstacles that had held them back in the fulfillment of reform goals. The first one was the misinterpretation of the reform ideas. One of the tenets of the reform emphasized that teachers should place attention on the process of learning and guide students to experience tangibly how knowledge was generated. In practice, due to teachers' superficial understanding of the tenet, their lessons were oftentimes embellished with what looked like many activities, but led to no meaningful outcomes in nature. In Teacher Me's words:

> Teachers seem to design fancy scenarios to engage students, and classrooms are indeed full of noise and giggles throughout the lesson. But, students oftentimes learn nothing at all. Half of the 40-minute lesson is simply putting on a show…not natural, not authentic learning.
> (Teacher Me, Interview #4)

She suggested that reform leaders and teachers also value the effectiveness of instruction. That is, what mathematics students learned out of one lesson should always be the focal point.

The second unaddressed issue was the relationship between the three-dimensional objectives promoted by the N.M.C.: Basic knowledge, basic skills, and emotion and attitudes about mathematics. Basic knowledge and basic skills were considered two major learning objectives of mathematics. Teacher Me stressed that it was unclear what constituted basic knowledge or basic skills in school mathematics and which was more important. More problematic was that basic skills could hardly be assessed via traditional paper-pencil tests.

Furthermore, the actual school conditions ran counter to reform advocacies. For instance, for classes with 60 students, Teacher Me questioned how hands-on activities could be carried out realistically. It was unfeasible logistically to use any station-based teaching strategies. In addition, the school could barely provide resources to support teachers and students. Principal Huang admitted that since even his office was not equipped with a computer, teachers could hardly integrate I.T. technologies into teaching and learning.

Making it worse, teachers were required to finish the prescribed instructional duties on time at the end of each semester. "If you could not finish teaching the whole volume of the textbook, you were causing an official teaching accident" (Teacher Me, Interview #4). Consequently, such teachers would be penalized in the year-end performance evaluation which might result in their being taken away from teaching main subject areas.

However, the real bottleneck of the reform implementation was the approach to assessment. Regarding the issue of overreliance on paper-pencil tests, Teacher Hua stated:

> The only standard of evaluation is test grades, by one single test paper. Singular assessment, pure test grades, then [students] have to learn pure

> knowledge, acquiring no skills. Pure knowledge, pure test problems. Because in the test students need to obtain as high test grades as they can, [we] have to be highly efficient in regular teaching…Without doing test practice papers, [students] don't know which problems they cannot solve and which knowledge areas they haven't mastered.
>
> (Teacher Hua, Interview #3)

In order to cope with the test, to varied extents, teachers had to adopt the "efficient" ways to prepare students for high test grades. Teacher Wen noted that if the standard of judging the success of students by test grades did not change, teachers would inevitably treat students unfairly. She revealed the common feelings of teachers:

> For students with good test grades, we teachers are fond of them genuinely. Even if the child might be not look nice, you feel he or she is adorable. For example, if two students are both violating school rules, you truly dislike the one with poor test grades. He or she simply has no capital to do that! This is the real feeling of teachers. Even worse in high school, high school teachers have the college entrance rate to evaluate. If your class has more students going to college, the teacher will be very honorable. If you have fewer, you feel shameful too. So, these poor students are the nails in teachers' eyes and thorns in teachers' flesh!
>
> (Teacher Wen, Interview #1)

The problem might be caused by a lack of effective means to assess the dimension of *abilities*, since "knowledge can be tested on a single piece of test paper, but abilities are invisible and cannot be gauged through paper-pencil exams" (Teacher Me, Interview #4). Teacher Me suggested that this predicament potentially could be addressed by experts in teacher professional development by offering teachers more creative approaches to assessing students' abilities.

Summary

Owing to the curriculum reform, teachers at Pioneer School experienced apparent instructional change. For example, teachers expanded their views of mathematics, and they knew a variety of instructional approaches, such as using small groups, conducting hands-on activities, and having life-relevant assignments. Professional development activities within the school and outside was significant to enable teachers' instructional change. Nevertheless, teachers' misinterpretation of reform ideas, overpopulated classes, and the exam-oriented culture collectively posed hurdles to the complete implementation of the curriculum reform.

Note

1 Ton might refer to: Long ton, a unit of weight equal to 2,240 lb avoirdupois (1,016.05 kg); a short ton, chiefly used in North America, is a unit of weight equal to 2,000 lb avoirdupois (907.19 kg); or a metric ton, a unit of weight equal to 2,205 lb (1,000 kg). In this case, the teacher meant metric ton.

10 Key findings

Initiation of the reform

The curriculum reform started unevenly at Merits School and Pioneer School. Merits School seemed to encounter the reform without much preparation, while Pioneer School prepared itself well in advance by undertaking a reform-aligned experiment.

Prior to the reform, Merits School was not carrying out any formal or systematic reform-aligned experiments and the reform was imposed by the D.E.B. Without much proper professional development, both administrators and teachers were left disoriented in the beginning. Even though administrators had the opportunity to attend training outside the City, and other teachers were exposed to reform theories during the summer workshops, such professional development was considered rather superficial.

Pioneer School, on the other hand, had been actively involved in a reform-aligned experiment well before the reform was started nationwide. Under the support of the City's educational authorities, the school initiated the I.E.M.E. Experiment, which paved the school's way for immersion in the national curriculum reform.

Implementation of the reform

The implementation of the reform took different shapes at Merits School and Pioneer School. The two schools responded to the reform differently. Pioneer School implemented the reform more authentically and consistently than Merits School.

Merits School adopted a two-faced strategy to cope with the reform: One face was to the outsiders, and the other was to the insiders. School administrators did not unanimously support the reform. Instead, they had conflicting views on it. As described previously, even though V.P. Yu attempted to promote reform pedagogies, T.D. Wang and the then school principal disagreed with her. Consequently, V.P. Yu had to compromise. Once the external policy environment no longer favored the reform, administrators in the school easily regained their consensus: Teaching for the Uniform Examination.

Teachers at Merits School were aware of the discordance of administrators regarding the reform. Cooperating with administrators, on occasions of lesson demonstration, competitions, and inspections, the school managed to put on reformer camouflage. Administrators intentionally displayed reform-based artifacts, teachers knowingly demonstrated reform-aligned lessons, and students behaved cooperatively in small groups and hands-on activities. As for day-to-day teaching in the school, teachers were teaching to tests, because reform pedagogy was considered more time-consuming and less effective in having students grasp the content and succeed in timed, paper-pencil examinations. Being evaluated by students' test grades, having overly large class sizes, and lacking resources became the most commonly expressed reasons for preventing teachers at Merits School from truly implementing the reform.

Pioneer School took the reform seriously. As shown in the data, administrators and teachers were willing to initiate educational experiments. Notably, ordinary teachers at Pioneer School had some say in the reform. For example, Teacher Wen, Teacher Quan, and particularly Teacher Me, were not passive implementers; throughout the I.E.M.E. Experiment and the reform, they participated in the process of decision-making, contributed their views, and engaged their colleagues in T.R. group-based professional development.

Pioneer School's reform encountered one major obstacle in 2006 because of a change to the school's status. The school was downgraded from being directly managed by the C.E.B. to being managed by the D.E.B. As a result, Pioneer School had greater external involvement from the D.E.B. and the District T.R. Center, whose directives they previously could have ignored. Pioneer School once had had the privilege to design and administer tests on its own, but now the school was no longer exceptional. It had to subject itself to the district-wide Uniform Examination. Their degree of autonomy was substantially compromised. When "all schools turn[ed] to focus on exams" (V.P. Yang, Interview #1), administrators at Pioneer School could not overlook test preparation either.

Although confronted with all these challenges, such as large class sizes, the lack of technologies, drop in school status, and the pressure of district Uniform Examination, the data indicate that teachers at Pioneer School demonstrated an upbeat morale overall. Teacher Me was confident that the school's long-standing culture – developing students' diverse abilities and intelligences – could not be easily manipulated. Teachers reported that they attempted to retain some reform practices even after the school was downgraded. For instance, they still occasionally engaged students in small group tasks, used manipulatives in instruction, and had students write mathematics journals and design mathematics posters as alternative approaches to assessment.

It needs to be mentioned here that student academic achievement at Merits School, measured by test grades, was much better than that at Pioneer School. Merits School ranked in the top three in all grade levels among all schools in the district, while Pioneer School ranked only 10th–15th . But, it would be wrong to think that Pioneer School's use of reform practices led to this gap. First, Merits School spent significantly more time teaching mathematics than Pioneer School. During the spring semester, Merits School had at least 13 periods of mathematics lessons per week, while Pioneer School only had 7 periods. Merits School almost neglected music, art, and physical education lessons throughout the spring semester, while Pioneer School held all classes to the last week of the semester. Second, Merits School had a much heavier workload than Pioneer School. Grade 6 students at Merits School were drilled with 131 sets of test practice papers from early March to early July, while Grade 6 students at Pioneer School only had 42 sets. Most likely, the achievement of high test grades and a top ranking at Merits School was at the cost of students' time, well-being, and all-round development.

Reforming teaching evaluation

At the organizational level, both Merits School and Pioneer School retooled the teaching evaluation system that included five major teaching activities: Lesson planning, instruction, student work, tutoring, and assessment, historically constituting the components of teaching norms in China (Liu & Teddlie, 2005). In particular, they infused reform ideas in those five teaching activities. Both schools stipulated detailed regulations to enforce reform-minded teaching evaluation. By closely attending to teachers' behaviors, theoretically, schools promoted reform-aligned teaching.

In reality, however, Merits School did not authentically implement the reform-aligned teaching norms. At Merits School, neither administrators nor ordinary teachers truly cared about conforming to teaching norms, since what really mattered to them were test grades. In this school world, test grades obtained a unique and singular significance. As illustrated above, test grades were the singular standard to judge the performance of the school and teachers at Merits School. To administrators, test grades concerned the school's competitive ability and survivability; to teachers, higher test grades suggested that they were better than their peers, thus preserving their professional dignity and saving face. Under this heavy test-centered culture, reform-minded assessment could hardly find its place at Merits School. To achieve outstanding test grades, all school people reached tacit consensus: Drilling students heavily.

Thus, teachers at Merits School chose to conduct the reform half-heartedly and to submit fabricated materials, and administrators chose not to make a fuss about it. The school ceremonially conducted inspections of teachers and produced a wide range of artifacts and records, such as the reform-minded curriculum schedule, lesson plans, and suggestions on alleviating and diversifying student work. But, both administrators and teachers were conscious that those products were crafted to cope with external inspectors and outsiders. In this way, Merits School successfully secured its two faces. Not only did it have an impressive record of reform activities, but also it maintained its superiority in district-wide Uniform Examination.

Pioneer School showed a different case from Merits School. Implementation and evaluation of teaching norms were treated seriously at Pioneer School. Pioneer School institutionalized a comprehensive teaching evaluation system that included teaching norms, behavior, school-based teaching research, meeting participation and completion of paperwork, parent review of teaching, and quality of teaching. More importantly, the school chose to enforce the system. The school had an immaculate record of regular inspections of teaching norms in conjunction with the other five aspects. At the end of the school year, every teacher in the school was scored and ranked according to those criteria.

Overall, the institutions of teaching evaluation attained different significance at Merits School and Pioneer School. At Merits School, what truly mattered to administrators was whether the school would rank number 1 in district-wide examinations, and to a teacher it was whether he or she could avoid ranking low in the grade group. Teaching norms became less meaningful, if not meaningless, at Merits School. Pioneer School did not have to excel beyond others in district tests, and teachers were under less pressure to pursue high test grades. The school adopted more plural standards to gauge and demonstrate the performance of the school itself and of teachers.

But, Pioneer School's reform-minded teaching evaluation system was not unquestionable. First, a regular teacher had to submit 19 kinds of written proof each semester, which certainly was a great deal and incurred the disapproval of teachers. It left one to wonder how teachers could possibly manage that amount of work without sacrificing teaching. Second, as for the quality of teaching category in the teaching evaluation system, the school relied primarily on unit and final test grades to judge a teacher's teaching performance. Paradoxically, such an evaluative measure ran against the reform ideas: Promoting alternative, reform-minded assessment approaches that would be most conducive to students' all-round development.

Teachers change and professional development

The study shows that teachers at Merits School and Pioneer School demonstrated positive attitudes towards the reform and looked forward to genuine change in mathematics teaching and learning. Teachers in both schools were seen having developed the skills to use reform pedagogy, such as small group-based cooperative learning to engage students in hands-on activities and to facilitate group or whole-class discussions. These positive signs are congruent with the findings in other studies (Lwe & Ni, 2011; Sargent, 2011). Teachers also reported that they tried alternative assessment approaches. For instance, at Merits School, Teacher Zhang and colleagues asked students to measure the perimeter of the playground and keep mathematics journals in the early years of the reform; at Pioneer School, Teacher Me still asked her students to design mathematics posters every semester.

However, whether teachers kept using reform pedagogy in everyday teaching was a totally different matter. At Merits School, the majority of teachers openly claimed that direct instruction and drills were the most effective and efficient means of securing high test grades. Reform pedagogy in this school seemed to be reserved for show times. Pioneer School was relatively better in that teachers sometimes still tried out reform approaches and school administrators openly supported such efforts, even though they had many challenges to face too.

Both Merits School and Pioneer School engaged teachers in rich modalities of professional development that took place at multiple levels. For instance, Merits School sent teachers to sister schools in Beijing, and Pioneer School had teachers learn in Tianjin. Merits School selected backbone teachers to actively participate in the district's 2-4-8 Project, and teachers at Pioneer School were guided by the City T.R. Office to conduct teaching research activities. Teachers at both schools spoke highly of those learning opportunities. In addition, each school also institutionalized the T.R. group-based weekly learning activity, and held school-wide teaching research event that lasted one month each semester.

In spite of all the virtues mentioned above, it was evident that teachers at Merits School did not wholeheartedly share with their colleagues in professional development activities. Rivalry among teachers was prevalent. Hiding knowledge from colleagues was the norm. Owing to its test-oriented culture, teachers at Merits School were less happy, disliked teaching, and developed more rivalry than collaborative relationships. They equated themselves with animals and menial workers. Competition among teachers rendered professional development pretentious and less useful.

In contrast, the majority of teachers at Pioneer School voiced a greater sense of satisfaction and belonging. Collaboration among teachers in the school was often seen. For example, experienced teachers like Teacher Me willingly coached early career teachers in formal and informal professional development events.

The role of parents in the reform

On the whole, both Merits School and Pioneer School ignored the role of parents. The parents were only involved marginally in all sorts of school activities, not to mention the reform. Merits School seemed not to have made systematic efforts to explain the reform to parents. Thus, most parents were not knowledgeable of the N.M.C. reform, had little understanding of the N.M.C., and were ignorant of what reform pedagogy was. They still viewed test grades as the only indicator of their children's success in school and their future lives. Ironically, parents' concern about test grades was alleged to be one of the main reasons why administrators and teachers in school stressed test preparation.

Administrators and teachers in Merits School showed dominant power over parents. Seldom did parents in this school stand up to challenge school administrators or teachers. If they did, they would have to apologize later or choose to transfer their children to another school.

Pioneer School was a bit better with regard to parental participation in the school. For instance, Principal Bao gathered parents and communicated to them the school's decision about the I.E.M.E. Experiment in late 1999. Yet, that was the only effort that the school made to inform parents of undertakings related to the curriculum reform.

Unlike Merits School, parents at Pioneer School were allowed to visit classrooms once every semester, observe teaching, and evaluate teachers at the end of each semester. Indeed, that reflected a higher degree of openness to parents of Pioneer School students. But, since parents had limited knowledge related to the curriculum reform, their observations and evaluation related more to teachers' diligence and conscientiousness in their work than to how well their teaching was aligned with the reform.

11 General discussion

As shown earlier, even though the centralized curriculum reform had been stressed and pushed forward determinedly as a national endeavor, local schools did not change accordingly. Both schools in this study did not fully implement the reform ideas as spelled out in the national plans. Nonetheless, the courses and the degrees of reform implementation of the two schools were different. One conspicuous difference was that Pioneer School seemed to have implemented the reform more faithfully than Merits School. Based on the findings, we conjecture that the interpretation power of local policymakers outside the school world and the power of school people collectively shaped reform implementation at the school level.

Beyond the school world: The interpretation power of local policymakers

It is known that implementation of reform policies is a complex process that involves active sensemaking of policymakers and implementers (Coburn, 2003; Cohen & Hill, 2000; Spillane, Reiser, & Gomez, 2006). Such sensemaking activities occur in the interpretation space between the policymakers who put forth the intended policies and these who interpret and enact the policies at different levels. That is, by dint of the interpretation power, educational authorities at each level not only implement policies imposed by the higher authorities but make policies of their own (Spillane, 1996). Policies intended at the higher policymaking level might be interpreted in a way that could possibly be incoherent with the original meaning.

In this study, the curriculum reform proposed by the national policymakers had to go through three or four layers of interpretation (between the M.O.E., the provincial education department, the C.E.B., the D.E.B., and the school administrators) before local schools could implement the interpreted policies. The findings suggest that the level of educational authorities that directly managed the school seemed to have the most influential interpretation power over it. In other words, which reform policies and how they should be implemented at the school level are highly dependent on highly dependent on the immediate educational authorities.

For Merits School, why the school adopted the two-faced strategy might be a result of responding to the reform policies that were interpreted and remade by the immediate educational authorities, the D.E.B. and the District T.R. Center. The messages that Merits School received and reacted to appeared inconsistent and conflicting with the national ones. Throughout the course of reform implementation, on the one hand, the D.E.B. and the District T.R. Center made various reform-oriented guidelines, developed reform-aligned new tools (e.g. the Comprehensive Assessment Plan), and provided institutional and material support for teacher learning (e.g. the 2-4-8 Project). On the other hand, the District T.R. Center reinstated the high-stakes examination and school ranking policy that ran counter to the national policy about assessment. Such high-stakes accountability measures essentially played the role of controlling the school, and its teachers and students (Brown et al., 2011), instead of promoting reform pedagogies. Merits School's close proximity to the local authorities seemed to leave it in a highly volatile policy environment. Whenever the local authorities came up with any new policies, either reform-oriented or old fashioned, the school had to respond accordingly.

The case of Pioneer School seemed to corroborate the above conjecture. Prior to 2006 when Pioneer School was in the jurisdiction of the city educational authorities, the policy interpretations of the C.E.B. and the City T.R. Office that the school received appeared more reform-aligned and consistent. The city educational authorities had created a relatively autonomous environment for Pioneer School, thus allowing it to cultivate a humanistic and innovative teaching culture and spearhead pioneering educational experiments for years. That might be why the whole school was able to engage in the new reform wholeheartedly. But, as the principal of Pioneer School changed, it seemed that reform implementation in the school started showing examination-oriented tendencies.

One lesson could be drawn from the above analysis. In the centralized educational system, it might be the case that aligning themselves with national policymakers was less of a concern for local schools that might care more about how their immediate supervisors viewed them. How a school responded to reform policies would by and large depend on how the school's immediate supervisors interpreted the intended policies. To ensure the success of the reform, reformers might need to consider narrowing the interpretation space between the highest policymakers and those who ultimately implement the reform, and decreasing the chance of policies being diluted.

Within the school world: The interpretation power of school people

It was seen that the interpretation power of local educational authorities exerted great external influences on how the two schools implemented the reform. Perhaps, schools have to adopt new policies symbolically in order to maintain legitimacy and survive in the local institutional environment (Meyer & Rowan, 2006). School people might take similar actions in order to survive in the profession of teaching.

Within the school world, school administrators and teachers actively interpreted the policies. They did not seem to place equal emphasis on all new policies that were imposed on them. This process of sensemaking and decision-making was much more complex than the commonly assumed willful distortion or resistance that leads to reform implementation failure (Cobb & Jackson, 2011). Implementation decisions seemed to be made based on delicate professional and personal considerations.

In the case of Merits School, school administrators did not naturally take a clear-cut stance: Implementing it or abandoning it, facilitating its implementation or obstructing it, and the like. Hypothetically, they could have pursued the reform genuinely and totally renovated its teaching culture, and focused on the all-round development of students. They also could have educated parents to develop a plural view of education, if parents were so obsessed with test grades. The actual implementation was the result of negotiations and game playing. For instance, every school administrator at Merits School had his or her own ideas about why and how the reform should be implemented (e.g. V.P. Yu and T.D. Wang's contention). They rationalized their arguments by drawing on prior experience (e.g. the district could not be trusted), school culture (e.g. the school ranking was too good to drop), and societal pressures (e.g. "parents looking at nothing but test grades" (T.D. Wang)). Their decision-making seemed to be predominated by personal motives that appeared mostly irrelevant to the intended purpose of the reform or the well-being of students, but more concerned with individual gains or loss such as job security and career advancement.

It appeared that the administrators identified what kind of policies could potentially benefit or harm them. Perhaps, their rich experience of coping with the higher policymakers allowed them to judge whether the new reform could succeed and be sustained, or if it would simply be another fad. If local policymakers were known for their arbitrary, unpredictable interpretations of reform policies ("changing three times in a day" (T.D. Wang)), and the effectiveness of reform pedagogy was uncertain, school people actually were in a state full of uncertainties. It would be understandable why administrators at Merits School arrived at the consensus of securing a top school ranking in the district examinations via the ocean of problems tactic. Top school ranking had always been the badge of Merits School's excellence to impress parents, attract students, and avoid school bankruptcy. Maintaining this image might be the first priority of the school leadership.

Teachers at Merits School were observed to be sensitive to what their school leaders valued and expected. Accordingly, they aligned their practices with such expectations. Like school administrators, they were clear that top school and personal ranking mattered to themselves, thus knowingly and cooperatively participating in the formation and maintenance of the two faces of the reform implementation. As found elsewhere, school leaders play a fundamental role to influence teachers in instructional change (Coburn & Russell, 2017; Spillane, 1996; Spillane, Reiser, & Gomez, 2006). The study revealed further how the perception of and response to the reform by administrators at Merits School delicately shaped teachers' decision-making.

Similarly, school administrators at Pioneer School did not passively or mechanically take in what the higher policymakers conveyed to them. They seemed to make implementation decisions based on careful deliberation too. Before 2006, Pioneer School had built the image of valuing educational innovations and the all-round development of students owing to its freer policy environment. It would have made little sense to school administrators and teachers to reorient their focus on high-stakes examinations that had not been their strength all along. It might explain why people in this school had unanimously participated in the I.E.M.E. Experiment and the reform that followed in earlier years.

After being downgraded in 2006, Pioneer School's approach to the reform showed a subtle but interesting shift: They not only kept carrying out the reform but also started paying more attention to test preparation. It seemed that people at Pioneer School arrived at a balance in practices: Not too pioneering and not too traditional. On the one hand, school administrators (e.g. Principal Huang and V.P. Yang) and teachers (e.g. Teacher Me) kept on highlighting the school's humanistic teaching culture and attempted to diminish intrusive influences from the D.E.B. and the District T.R. Center. This action might be out of the school people's genuine pride in and beliefs about the school's culture. Also likely is that they intended to maintain the school's special selling point and to tell it apart from other competitors in the district. On the other hand, school administrators and teachers placed a certain amount of attention to test preparation. It looked like a purposeful move in order to protect the school from being diminished in the district's high-stakes accountability appraisal, since low school ranking in the examinations would very likely tarnish its reputation.

One final conjecture is made to conclude this section. The two cases suggest that school administrators more likely chose to conform to a school's established teaching culture rather than change their own mindset. In other words, schools shaped school people rather than the other way around. The fact that Principal Yong became examination-oriented and Principal Huang was reform-minded might be more accidental. Maybe, if Principal Huang had been assigned to Merits School, he might have taken similar actions to the ones Principal Yong took.

Teaching evaluation and teachers as paralyzed professionals

Utilizing teaching norms to evaluate teachers is an old tool in China (Liu & Teddie, 2005). The reform infused the old tool with new contents and turned it into a new tool essential to successful reform implementation (Cobb & Jackson, 2011). In the hope of transforming teachers' practices, both Merits School and Pioneer School institutionalized new teaching norms at the beginning of the reform. The new schemes included preparing reform-aligned lessons, using reform-minded instruction observation protocols, developing diversified student work with enhanced cognitive challenges, pluralizing assessment approaches, and the like.

As described earlier, in reality, Merits School almost completely gave up most of the reform elements of the new teaching norms, and Pioneer School did not fully apply the new teaching norms either. Beyond the applicability of the new teaching norms at each school, a deeper issue was common to both schools.

Teachers are believed to play a central role in the enactment of reform initiatives (Cobb, Yackel, & Wood, 1992) and teachers are often regarded as professional agents capable of making changes (Cohen & Hill, 2000). Nonetheless, based on the findings of the study, one question has to be raised: Did teachers at both schools have the autonomy to make instructional decisions based on their professional judgment? Lortie (1975) probably best depicts the image of teachers' autonomy: Close the door and do whatever they want to do. That is, whether teachers hold power and control over workplace decisions distinguishes teaching from other lines of work (Ingersoll, 2011). It was clear that teachers in the two schools lacked professional autonomy. At best, those teachers were paralyzed professionals.

The data shows that teachers at Merits School and Pioneer School were tightly controlled by administrators inside and outside schools. Their teaching schedules were prearranged by the District T.R. Center or the City T.R. Office, curricular contents were standardized, and teaching progress was monitored. By dint of the teaching norms system, teachers' behaviors, such as how to dress, stand, speak, or grade student work, were clearly defined and closely monitored. Outsiders' intrusion into classrooms further deprived teachers of an already scarce sense of autonomy. Classrooms could be entered without advance notice and lessons observed by administrators, colleagues, inspectors, and parents. Taking parents' evaluation into account in teaching evaluation, as Pioneer School did, might further push teachers toward examination-oriented teaching, since most parents still held the notion that educational quality was bound up with test grades.

More problematically, evaluating and ranking teachers based on students' test grades were detrimental to teachers, which was particularly at issue at Merits School. Two consequences were notable. First, it further worsened teachers' sense of professionalism and left them in a vulnerable and insecure state because teachers' relative ranking could be so volatile that their sense of competence and confidence underwent constant self-doubt and challenge. Second, relationships between teachers were estranged. The fear of being overshadowed by colleagues compelled teachers to join in the race for high test grades and alienated them by making them into rivals. Beating down colleagues in high-stakes examinations seemed the only way to preserve or redeem one's self esteem.

This study provided further evidence that strong accountability pressures from the system and society in general significantly reduced the possibility of teachers' change (Brown et al., 2011). Coburn (2003) observed,"[B]ecause classrooms are situated in and inextricably linked to the broader school and system, teachers are better able to sustain change when there are mechanisms in place at multiple levels of the system to support their efforts" (p.6). Understandably, the reform would be less likely to succeed if what teachers received was not support but constant surveillance and intrusions which disregarded them as professionals.

Evaluative professional development

Merits School and Pioneer School in this study both provided high quality professional development schemes that offered teachers a variety of learning opportunities to grow. Those activities also had teacher evaluation functions. Three strengths were notable. First, a wide range of professional development activities assisted teachers to build up professional learning communities in which they could "interact and collaborate regularly around issues of teaching and learning, and engage in the production and consumption of knowledge about improved practices for student learning" (Sargent & Hannum, 2009, p.258). Such activities as collective lesson planning, peer observation, lesson demonstration, and lesson narration provided the ground for both experienced teachers and novice teachers to interact with each other and develop a common set of reform-based practices. The regularity and multiplicity of professional development at both schools could enable teachers to foster collegial relationships, nurture enriching learning communities, and grow on the job (Fullan, 1991).

Second, professional development that took place at multiple levels extended teachers' social networks in both schools (Coburn & Russell, 2017). For instance, at the school level, Pioneer School had experts from the City T.R. Office to visit the school and coach teachers in teaching research. At the local level, backbone teachers at Merits School had many chances to visit other schools and learn with the best teachers via the district's 2-4-8 Project. At the provincial and national level, both schools regularly had teachers visit sister schools in more economically developed provinces like Beijing and Tianjin. Such professional development arrangements that were multitier and content-oriented allowed teachers to get much greater access to external experts who increased teachers' tie span and broadened their scope of expertise. Also, frequent and meaningful interactions with such experts might enhance teachers' in-depth understanding of reform curriculum and pedagogy (Cohen & Hill, 2001).

Third, professional development at both schools promoted job-embedded learning. On-the-job learning is believed to significantly promote teachers' instructional change (Parise & Spillane, 2010). Both schools held regular T.R. group-based or school-wide teaching research events. These events provided teachers the venue to interact with one another and promote learning onsite. In short, well established professional development helped cultivate professional learning communities, broadened teachers' social networks, and enhanced on-the-job learning at both schools.

However, professional development alone, no matter how high quality or systemic, did not naturally lead to successful reform implementation or teachers' permanent change. In other words, high quality professional development is necessary but far from sufficient to develop and sustain reform practices in schools. Teachers cannot be transformed without tackling the core issues that obstruct school administrators and schools from genuine reform implementation.

Parents' involvement in the reform

As said earlier, parents had no substantive involvement in the reform at either school, but the culture of how parents were treated showed a difference between the two schools. Merits School seemed to be more aloof and dominant in interacting with parents than Pioneer School. Unlike Merits School, Pioneer School created opportunities for parents to participate in school affairs, and allowed parents to visit classrooms and observe teaching once every semester. One possible explanation was that Pioneer School had developed a relatively open and democratic culture. Yet, another possibility was worth considering. As mentioned earlier, many parents who had higher social status and stronger *guan xi* (connections) selected Pioneer School for their children. School people tended to be very careful about their ways of working with students and parents. The latter might substantiate the point that parents of lower socioeconomic and educational background are more disadvantaged in mathematics education reform (Remillard & Jackson, 2006), and probably in education in general.

At both schools, parents were considered the culprit of the examination-obsessed culture because of their ignorance of education. On the one hand, this claim could be school people trying to legitimize their examination-oriented practices and free themselves from the professional guilt. On the other hand, parents might really believe in test grades. As the study shows, fears were prevalent among parents (some were teachers themselves): They dreaded to think that their children would be spun off the test mill and lose higher education opportunities that would secure them a good quality of life.

This phenomenon seemed to reflect a deeper educational issue: The purpose of education was displaced. People placed test grades first, not other developmental needs of students. Parents, teachers, principals, and students all centered on this belief that mathematics learning was all about test scores. They were the invisible force to perpetuate this examination culture, and at the same time they were the victims of their own fears. Thus, to involve parents equally, meaningfully, and respectfully in the reform might necessitate a fundamental change of school culture and societal values.

12 Concluding remarks

To what degree did the constructivist curriculum reform get implemented in the two schools? To answer this question, let's revisit the tenets of the N.M.C. as outlined by the N.M.C. Standards (2001), juxtapose them with the findings, and consider if the tenets were embodied in any form at the two schools. The first tenet states that the school mathematics curriculum should "focus on fundamentality, universality, and developmentality, and be accessible to all students" (M.O.E., 2001, para. 3). This principle is particularly concerned with and embodied in curriculum materials, mainly textbooks (Lwe & Ni, 2011). Even though the reform proposed to decentralize the management of curriculum and advocated school-based curriculum, governance of the mathematics curriculum was at a local level. A few series of textbooks that were endorsed by M.O.E. and composed on the base of fundamentality, universality, developmentality, and accessibility (Lu & Wang, 2004), were available to schools. Both schools adopted such reform-minded textbooks as the first step of reform implementation.

The second tenet concerns what mathematics is and what mathematics functions as. However, this tenet was not fully enacted. School mathematics is often considered value-free, culture-independent, and objective. Frankenstein (1992) pinpoints the significance of contextualizing mathematics and locating mathematics education in the culture and history of a society. But, this seemed not to be the case in this study. The results of the study suggest that little of the Chinese traditional and indigenous mathematics culture was seen in the mathematics materials provided. As administrators and teachers implied from time to time, they viewed Western countries, specifically the U.S., as the model to follow and the source for of inspiration and validation of their practices.

Another issue, as the findings reveal, is that mathematics was predominately regarded as an academic subject that mattered to students' success of schooling. Those who performed outstandingly in high-stakes examinations were awarded higher quality educational resources, including special treatment like specially assigned teachers, smaller class sizes, and extra attention. Primarily, mathematics became a sorting and tracking scheme to stratify students. As a result, mathematics was turned into a tool of "linking (perceived) mathematics ability to intelligence, and thus to power and privilege" (Stinson & Bullock, 2012) instead of a useful tool to benefit life, work and learning.

The third and fourth tenets are at the heart of the reform. They set the framework for mathematics content, learning, and teaching. Accordingly, inquiry-based, cooperative learning is advocated to approach realistic, meaningful, and challenging mathematics content that meets students' diverse learning needs. Students were expected to have the ownership of learning with teachers' working as facilitators, provocateurs, and questioners (Fosnot, 2005).

The findings indicate that most teachers' beliefs had changed and were in line with these reform principles. They could state clearly the importance of respecting students' agency in mathematics learning. They also recognized that teachers should take a step back and create opportunities for students to explore on their own.

As for change of instructional practices, teachers did demonstrate at certain points in time that they engaged students in hands-on activities and work in small groups. To some degree, such practices were conducive to the development of students' sophisticated mathematical thinking and reasoning, and to enhancing positive interactions between students. But, a critical look at the study indicates that inquiry-based teaching and learning appeared more contrived than natural. This observation applied to both schools, even though teachers at Pioneer School seemed more prone to engaging students in hands-on activities and small group discussions more often than their counterparts at Merits School.

The major concern was that the activities assigned to students tended to be examples selected deliberately by teachers from the textbooks, the unambiguous authority. Throughout the process, teachers worked essentially as directors defined by the textbooks, setting the scene, composing the plot, fixing specific goalposts, and maneuvering students to reach the goalposts. Students seemed to cooperate knowingly with teachers and oftentimes arrived at the desired answers that were pinpointed in the textbooks. More precisely, students were passive performers and followers rather than active agents who could explore the topics from a multiplicity of angles or in greater depth and scope. The inquiry was thus constrained in case "students get out of control" in lessons. If students were already conscious of what they were asked to look for and how, probably they were not truly inquiring.

The fifth tenet emphasizes that assessment of student learning should be plural, ongoing, and equally attending to students' mathematical proficiency and emotional values. The study revealed the apparent misalignment between assessment approaches and the reform curriculum and pedagogies. Policymakers at different levels seemed to have no capacity to invent innovative assessment approaches. Local practitioners were left to return to the traditional paper-pencil based means of assessment. Relying almost solely on high-stakes tests and examinations led to a number of negative consequences, particularly the displacement of educational goals and the chain of tortuous vicious circle among administrators, parents, teachers, and students.

The last tenet envisions the integration of I.T. in mathematics education. The study suggests that this tenet was partly exemplified in instruction. Sophisticated I.T., such as presentation software and animations, was used to represent mathematical concepts and operations. Yet, since teachers' everyday instruction remained slightly changed, employing I.T. in instruction seemed to be discredited as "tricks" that purported to impress visitors during "show moments".

The review of the intended tenets and those enacted in the constructivist curriculum reform in the two schools suggests to us that the reform implementation did not live up to most of the visions that the reformers set forth. Although one school might more sincerely and persistently devote itself to reform implementation than the other, and some teachers might persevere more in trying out reform ideas than others, overall, the reform seemed unsuccessful in most aspects. Similar to findings in other studies (Cohen & Hill, 2000; Spillane, 1999), the study indicates that the curriculum reform seemed to be assimilated to old values and practices – a mixing of disconnection with students' real lives, inequitable use of mathematics for social stratification, contrived mathematical activities, and high-stakes approaches to assessment. Even worse, it seemed that the ideas put forward in 1984 – "alleviating the workload of children," "developing all-round children," "respecting students' ownership of learning," and "teachers are the leading facilitators" – were not yet fulfilled (Merits School, 1984). There might be a need to pause, carefully plan, and reenergize the current reform.

Centralized reform and conflicting policies

This study provides empirical evidence that even in a highly centralized education system, curriculum reform cannot succeed simply because reformers have the power to summon enormous political, economic and human resources and can impose the reform from the top level to the school level as a national mandate. Put it in another way, implementation of a centralized curriculum depends highly on sensemaking and decision-making at a local level, which can yield considerable variation between schools. Successful implementation at the local level is premised on at least one critical condition: Coherent and consistent policies.

The current curriculum reform in China was ambitious in terms of its scope and depth, but, a reform touted as groundbreaking and thorough might be doomed to fail if it lacks consistency and coherence in policymaking. Policymakers tend to shift their policymaking. This was what Merits School experienced: Qualities-oriented education, creative education, innovative education, modern mathematics, standards-based curriculum reform... in less than a decade. The faddism for something new is so prevalent in education that teachers may simply get overwhelmed and disoriented. In this sense, teachers' so-called "practicality ethic" (Doyle & Ponder, 1977–1978) is teachers' calculated response to the fluctuation of policymaking and an attempt to get reoriented in the face of multiple strands of ideas. "Sticking to the one way to cope with tens of thousands of changes" (Teacher Fu) was thus not only a handle for teachers to grasp in the turbulence of changes, but also a constructed state of emotion: Nonchalance to change anyway because of disillusion. Of course, the worst backlash of proposing educational innovation and aborting it halfway is that reformers might run out of credit: "Whoever believes in them will be fools" (Teacher Su).

Professional development: Necessary but insufficient

High quality professional development is a central component in nearly every modern proposal for improving education (Guskey, 1999). The multitier professional development system in China seems very effective to pass on reform ideas at scale. The system is open to new ideas and efficient in spreading them out. It can successfully situate teachers' learning in the workplace and nurture a professional learning community, and at the same time effectively extend teachers' networks.

Of course, having this system in place can successfully improve teachers' knowledge and change their practices, but such change cannot be sustained. Teachers' permanent change depends on an array of other factors that are more determining. One major factor is standards of teaching evaluation at the local and school levels. The link between students' test grades to the quality of schools, administrators, and eventually teachers could only force teachers to relinquish most reform pedagogy that they learned in professional development activities for the sake of efficiency and effectiveness in test preparation.

Thus, this situation results in a grotesque conflict between reform-oriented professional development and teachers' day-to-day examination-oriented reality. What they learn and demonstrate in professional development activities are perfectly reform-minded, but what they practice in classrooms is totally different. This inconsistency between what is being preached and what is actually used may be so unsettling that teachers doubt the value of professional development and give up learning. Indeed, why bother "putting on a show?"

Parents: The oppressed oppressors

Curriculum reform is not simply an endeavor constrained to the sphere of schools. It is deeply embedded in and defined by the social, cultural, and economic reality of society. Policymakers cannot overlook these factors in designing and implementing policies as if the reform would happen in a value-free vacuum. Parents' influences, apparently overlooked in this case, can be powerful.

The rich feelings of insecurity, worries, bewilderment, helplessness, and even anger shown by parents and teachers, who might be parents themselves, suggest that schools carry most of their hopes for their children. They seem to be victims of the examination-obsessed education. Yet, they must also be active participants in creating this misery.

Parents may have to make sure that schools are not failing their children according to whatever standards are acceptable in broader society. Without seeing other alternatives, what parents want is that their children can achieve higher, faster, and stronger test results than their fellow competitors in this Olympics-style education. Schools may have to respond to this call while in the meantime taking advantage of the motives of parents and justifying their action of working students hard.

In the most important school businesses, including the implementation of reforms, parents are kept out of the reform equation or are only marginally involved. Rather than blaming parents for their obsession with test grades, one thing that reformers and school people should do and can do right away might be to educate parents and update their understandings of education.

Implications of the study

Implications for policymakers

Three implications are made. First, it is suggested that large-scale implementation of a new curriculum may achieve desired outcomes in instructional change at a wide level (Ball, 1990). Yet, this study suggests that centralizing curriculum and implementing it on a national scale cannot naturally lead to good practices on the teachers' side. Policymakers need to single out and tackle the central factors that influence teachers' professional behaviors. In this study, it is the misalignment of reform curriculum and pedagogy with school evaluation at the local D.E.B./District T.R. Center that broke the reform apart.

Recent years have seen many countries turn towards developing national mathematics curriculum standards and curricula. For instance, the U.S. has been seen moving toward this direction (Common Core State Standards Initiative, 2010). The efforts to centralize curriculum control and management might not directly lead to changing teachers' practices and improving students' academic achievement. Clearly, it does bring many apparent benefits. Different regions in the country can have shared knowledge and practices. Student assessment outcomes are more comparable. People do not have to waste resources on repetitive things. However, implementation of the curriculum is a highly localized and decentralized act. Local districts and schools tend to determine where the reform goes (Ogawa et al., 2003), depending on the local cultural legacies, economic conditions, and administrative capacities.

Second, policymakers should be especially careful about being consistent in their policymaking, being careful about reform ideas and not attempting to introduce new ideas too frequently, and not wearing out teachers' patience with policy flip-flops. Moreover, policies that create rivalry among teachers and between schools should best be avoided. In particular, high-stakes accountability pressures on schools and teachers have many detrimental effects on school people and students. As the U.S. increasingly calls for more rigorous accountability measures on schools and especially teachers, people should be prepared to see that teachers' moral will be lowered, teachers' professional autonomy intruded, and teachers will limit themselves to tested content areas.

Third, to change teaching practices in schools, teachers might not be the right target, at least not always, of educational reform. Teachers tend to be the ones that policymakers easily target and the ones that are held accountable to reform outcomes, since it is widely believed that teachers are the key arbiters of instructional content and practices (Cohen & Hill, 2000). This assumption might not be so true in a centralized education system, even less so in a highly decentralized system like the U.S. (Ingersoll, 2011). In school organizations, teachers may have no power to decide what and how to teach in their classrooms even if they are perfectly knowledgeable of reform practices. Policymakers need to carefully examine the parameters of the local education administration and schools, and zero in on these organizational or institutional factors that impede teachers' transformation. They, not teachers, should be the first to change.

Implications for teaching evaluation

Teaching quality is the major concern of educational administrations and the general public. Understandably, the quality of teaching has a direct impact on students' academic achievement. Increasing policymaking efforts are seen as tightening grips on teachers. Associating teacher performance to students' high-stakes test scores becomes more popular. Policymakers have to be very careful about the negative consequences caused by such policies. Teachers may be forced to distort the purpose of education and teach to test, as Merits School did in the study.

One alternative approach to ensuring the performance of teachers is school-based teaching evaluation. If used properly, it is an effective, data-driven instrument and can provide timely feedback to school administrators and teachers themselves about the quality of teaching, but, the system has to be used for the authentic needs of teachers to encourage and improve reform practices. It can be counterproductive and perpetuate undesired practices if the school-based evaluation looks for one thing while external accountability policies pressure teachers to do another.

Implications for professional development

Rich modalities of professional development programs established at multiple levels are positive to teachers' learning. This is very similar to the Japanese model of Lesson Study (Yoshida, 1999). Particularly, the teaching research group within schools can be utilized to disseminate new curriculum and pedagogy, and to develop shared teaching strategies. However, it should be noted that such professional development schemes are built on Chinese culture.

In China, a Chinese individual is enmeshed in interlocking reciprocal relations with others and becomes a shared significant other (Sun, 1991). In such a collectivist culture, teachers tend to sacrifice themselves for others. Whether professional development models growing in this culture can take root in more individualist-oriented cultures, like the U.S., needs further exploration.

Appendix 1: A chronology of the events of the I.E.M.E. Experiment at Pioneer School

Table A1 A chronology of the events of the I.E.M.E. Experiment at Pioneer School

Date	*Location*	*People*	*Form*	*Key points*
10/12/1999	Teaching Supervision Office	Director Zhi, Teacher Me, Teacher Chun, Teacher Xiang, Teacher Jun	Team meeting	(1) Learned materials concerning I.E.M.E. Experiment. (2) Director Zhi assigned tasks to teachers to prepare for the Experiment.
11/06/1999	The Auditorium of another school	Principal Bao, Fellow Xu, experiment teachers, and parents of experimental students	Parents' conference	(1) Fellow Xu elaborated on the theoretical foundations, characteristics, and goals of the Experiment. (2) Principal Bao discussed with parents how teachers and parents could collaborate in the Experiment.
3/17–3/18/2000	Computer Classroom	Fellow Xu, Director Zhi, Teacher Jun, Teacher Xiang, Teacher Chun, Teacher Me	Research lesson study	(1) Observed Teachers Xiang, Me, and Chun's lessons. (2) Teachers engaged in post-lesson discussions. (3) Fellow Xu provided in-depth critique and suggestions.
04/20/2000	Vocational High School	All experiment teachers	Training on the N.M.C. Standards	(1) Teacher Rong of No. 5 Elementary School gave open lessons. (2) Studied the N.M.C. Standards thoroughly.
06/24/2000	The Auditorium of another school	Principal Bao, Director Zhi, all experiment teachers	Open lesson study	(1) Super-senior Teacher[a] Cheng from Beijing demonstrated one open lesson. (2) Attended the workshop on I.E.M.E.

Table A1 Cont.

Date	*Location*	*People*	*Form*	*Key points*
09/14/ 2000	Computer Classroom	Fellow Xu, Director Zhi Teacher Jian, Teacher Shu, Teacher Ling, Teacher Quan	Research lesson study	(1) Teachers Quan, Ling, and Shu gave research lessons. (2) Fellow Xu gave academic guidance. (3) Teachers held in-depth discussions.
09/28/ 2000	Conference Room	Principal Bao, V.P. Yang Director Zhi, Teacher Jun, Teacher Xiang Teacher Lan, Teacher Hao, Teacher Jian, Teacher Shu, Teacher Ling, Teacher Quan	Meeting	(1) Director Zhi announced the plan of the I.E.M.E. Experiment for the 2000–2001 school year. (2) V.P. Yang elaborated on how to conduct teaching research. (3) Principal Bao reviewed the progress of the experiment made last semester and stated the goals of this semester.
11/06/ 2000	Computer Classroom	Principal Bao, Director Zhi, Teacher Quan and others	Activity lesson in math	(1) Teacher Quan taught an activity lesson in math. (2) Director Zhi introduced the features of and strategies used in an activity lesson. (3) Principal Bao critiqued the activity lesson.
05/16–05/18/ 2001	Computer Classroom	All experiment teachers	Theme lesson study	(1) Teachers Xiang, Hao, Lan gave research lessons. (2) Researched to establish the I.E.M.E. Instruction Model.
06/12/ 2001	Computer Classroom	Fellow Ruo from the province, Fellow Xu, Principal Bao, all experiment teachers	Lesson study	(1) Teachers Chun, Me, and Xiang demonstrated lessons. (2) Fellow Xu from the City T.R. Office critiqued the lessons. (3) Fellow Ruo from the Province T.R. Office offered academic guidance. (4) Teachers asked questions to the experts.

(*continued*)

Table A1 Cont.

Date	*Location*	*People*	*Form*	*Key points*
07/06/ 2001	Conference Room	Principal Bao, V.P. Yang Director Zhi, all math teachers	Sharing of experience	(1) Experiment teachers shared learning. (2) Director Zhi summarized the work to date. (3) Principal Bao concluded the meeting.
09/24/ 2001	Conference Room	Principal Bao, Director Zhi, experiment teachers	Research projects meeting	Put forward the research plan for the school year.
09/20/ 2001	Conference Room	All experiment teachers	Seminar	(1) Established the student assessment scheme for the I.E.M.E. Experiment. (2) Chose the specific content area to assess for each experimental class. (3) Decided to assess students from December.
12/12/ 2001	Computer Classroom	Fellow Xu, Director Zhi, Teacher Xiang, Teacher Jian, and others	Research lesson study	(1) Teachers Jian and Xiang gave research lessons. (2) Fellow Xu guided the I.E.M.E. Instructional Model. (3) Director Zhi proposed the model to evaluate students.
03/04/ 2002	Conference Room	Principal Bao, Director Zhi, Teacher Jun, Teacher Shu, Teacher Xiang, Teacher Jian, and others	Meeting	(1) Summarized what has been achieved to date (2) Formalized the I.E.ME. Instructional Model. (3) Made suggestions for future work.
04/18/ 2002	Computer Classroom	Fellow Xu, Director Zhi, Teacher Shu, Teacher Xiang, Teacher Jian, Teacher Lan, Teacher Quan, Teacher Hao Teacher Ling, Teacher Jun, Teacher Me, Teacher Wen	Research lesson study	(1) Teachers Quan, Lan, and Hao did research lessons. (2) Fellow Xu commented on the lessons and proposed the tasks for the next step. (3) Experiment teachers ask questions. (4) Director Zhi concluded the meeting.

Table A1 Cont.

Date	*Location*	*People*	*Form*	*Key points*
05/2002	Tianjin Experimental Elementary Tianjin Affiliated Elementary	Director Zhi, Teacher Jian, Teacher Quan, Teacher Hao, Teacher Lan, Teacher Xiang, Teacher Shu, Teacher Wen	External learning	(1) Head of the Education Bureau of the H District in Tianjin introduced how the I.E.M.E. Experiment had been carried out. (2) Observed lessons. (3) Exchanged ideas with the local teachers.
06/04/ 2002	Conference Room	Principal Bao and all experiment teachers	Post-learning meeting	(1) Teachers reported what they had learned in Tianjin. (2) Principal Bao raised questions and concluded the meeting.
10/18/ 2002	Computer Classroom	Fellow Xu, Teacher Ying and all experiment teachers	Research lesson study	(1) Teachers Hao, Quan, and Xiang conducted research lessons. (2) Revised and improved the I.E.M.E. Instructional Model. (3) Fellow Xu offered academic guidance.
11/25/ 2002	Principal's Office	Principal Bao, Fellow Xu, Director Zhi	Leaders' meeting	Discussed to select teachers to give open lessons in the City.
12/03/ 2002	Vocational High School	All city-level backbone teachers and experiment teachers	Research lesson study	(1) Teachers Lan, Quan and Chun demonstrated reform lessons. (2) Fellow Xu critiqued the lessons and provided training on the N.M.C. Standards. (3) Director Zhi shared Pioneer School's experience in carrying out the I.E.M.E. Experiment.
03/08/ 2003	Conference Room	All experiment teachers	Meeting	(1) Studied the advanced experience of other cities. (2) Decided the key themes of this year's experiment.

(*continued*)

Table A1 Cont.

Date	*Location*	*People*	*Form*	*Key points*
04/06/ 2003	Computer Classroom	Principal Bao, V.P. Yang and all experiment teachers	Theme lesson study	(1) Focused on the special issue of how to conduct small group-based, cooperative learning in the context of the Experiment. (2) Teachers critiqued each other's teaching and learned from observing and critiquing.
04/24/ 2003	Han County	Director Zhi and Teacher Quan	City-level teaching competition	Youth teacher Quan won first place in the City-level Teaching Competition for Youth Teachers.
09/26/ 2003	Principal's Office	Principal Bao, V.P. Yang, Director Zhi	Meeting	Discussed issues related to the final evaluation of the Experiment.
10/2003	Conference Room	All experiment teachers	Theme lesson study	(1) All experiment teachers taught one research lesson. (2) Teachers observed and critiqued each other's lesson. (3) Assessed students via test papers and observations, and prepared for the final evaluation

Note: Translated by the author from Pioneer School (2003b, pp.207–218).

[a] Super-senior teacher (*te ji jiao shi*; *te*=special, *ji*=level, *jiao shi*=teacher): In China, teachers may attain different professional ranks (*zhi cheng*; *zhi*=profession, *cheng*=title) on the basis of factors such as years of teaching experience, teaching performance, achievement in teaching research and so on. Each year, teachers need to file an application for the promotion of professional rank through education bureaus. At the elementary level, teachers' professional ranks consist of Elementary Level 1, Elementary Level 2 (equivalent to Middle School Level 1), Elementary Senior (equivalent to Middle School Intermediate), and Elementary Ultra-senior (equivalent to Middle School Senior).

Appendix 2: Test scores at Merits School and Pioneer School

Table A2 Average scores of Grades 1, 2, 4 and 5 at Merits School and Pioneer School in the July district-wide Uniform Examination

	Merits School			*Pioneer School*	
	Class	*Class size*	*Average*	*Class size*	*Average*
Grade 1	1	64	95.94	64	91.98
	2	63	95.72	64	91.26
	3	63	95.38	65	91.09
	4	63	94.37	63	90.74
	5	65	94.25	59	90.25
	6	64	93.44	64	86.92
		Average	**94.85**		**90.05**
	District average: 90.78				
Grade 2	1	62	98.57	59	97.47
	2	61	98.20	56	97.25
	3	59	97.88	58	95.84
	4	58	97.54	58	95.78
	5	61	96.68	58	94.79
	6	59	96.68	56	94.60
		Average	**97.60**		**95.96**
	District average: 95.01				
Grade 4	1	57	95.25	61	89.00
	2	58	94.73	57	88.85
	3	61	94.69	56	87.19
	4	56	94.12	61	85.63
	5	60	94.02	60	85.52
	6	61	93.37	56	85.30
		Average	**94.37**		**86.92**
	District average: 87.60				
Grade 5	1	51	97.12	67	89.22
	2	51	97.04	62	87.80
	3	54	97.02	63	87.60
	4	52	96.63	62	87.25
	5	52	96.35	65	82.70
	6	47	93.50	N/A	N/A
		Average	**94.28**		**86.90**
	District average: 90.92				

References

Agrey, L. (2004). The pressure cooker in education: Standardized assessment and high-stakes. *Canadian Social Studies, 38*(3). Retrieved on November 18, 2019 from https://files.eric.ed.gov/fulltext/EJ1073917.pdf.

Asia Society. (2005, December). *Education in China: Lessons for U.S. educators.* New York: Asia Society.

Ball, D. L. (1990). The mathematical understandings that prospective teachers bring to teacher education. *The Elementary School Journal, 90*(4), 449–466.

Becker, H. S. (1996). The epistemology of qualitative research. In R. Jessor, A. Colby, & R. Shweder (Eds.), *Essays on ethnography and human development* (pp.53–70). Chicago, IL: University of Chicago Press.

Beijing Normal University. (1999). *The 9-year compulsory education curricula-mathematics.* Beijing: Beijing Normal University Press.

Bishop, A. J. (1994). Cultural conflicts in mathematics education: Developing a research agenda. *For the Learning of Mathematics, 14*(2), 15–18.

Blumer, H. (1969). *Symbolic interactionism: Perspective and method.* Berkeley, CA: University of California Press.

Brooks, J. G., & Brooks, M. G. (1993). *In search of understanding: The case for constructivist classrooms.* Alexandria, VA: Association for Supervision and Curriculum Development.

Brown, G. T. L., Hui, S. K. F., Yu, F. W. M., & Kennedy, K. J. (2011). Teachers' conceptions of assessment in Chinese contexts: A tripartite model of accountability, improvement, and irrelevance. *International Journal of Educational Research, 50*(5–6), 307–320.

Cai, J. (2004). Why do U.S. and Chinese students think differently in mathematical problem solving?: Impact of early algebra learning and teachers' beliefs. *The Journal of Mathematical Behavior, 23*(2), 135–167.

C.E.B. (2008, April 9). *Opinions on the management of teaching in elementary and secondary schools in the City.* Internal report: Unpublished.

Charmaz, K., & Belgrave L. L. (2001). Qualitative interviewing and grounded theory analysis. In J. F. Gubrium & J. A. Holstein (Eds.), *Handbook of interview research: Context and method* (pp.675–694). Thousand Oaks, CA: Sage Publications.

Chen, Y. G. (1999, May). Tradition and innovation in the Chinese school curriculum. *Research in Education, 61*(1), 16–28.

China National Knowledge Infrastructure. (2019). *Educational statistics of China.* Retrieved on October 16, 2020 from https://data.cnki.net/trade/Yearbook/Single/N2021020052?z=Z017.

Cobb, P. (1994). Where is the mind? Constructivist and sociocultural perspectives on mathematical development. *Educational Researcher, 23*(7), 13–20.
Cobb, P., & Jackson, K. J. (2011). Towards an empirically grounded theory of action for improving the quality of mathematics teaching at scale. *Mathematics Teacher Education and Development, 13*(1), 6–33.
Cobb, P., Yackel, E., & Wood, T. (1992). A constructivist alternative to the representational view of mind in mathematics education. *Journal for Research in Mathematics Education, 23*(1), 2–33.
Coburn, C. E. (2003). Rethinking scale: Moving beyond numbers to deep and lasting change. *Educational Researcher, 32*(6), 3–12.
Coburn, C. E. & Russell, J. L. (2017). District policy and teachers' social networks. *Educational Evaluation and Policy Analysis, 30*(3), 203–235.
Cohen, D. K., & Ball, D. L. (1990). Policy and practice: An overview. *Educational Evaluation and Policy Analysis, 12*(3), 233–239.
Cohen, D. K., & Hill, H. C. (2000). Instructional policy and classroom performance: The mathematics reform in California. *Teachers College Record, 102*(2), 294–343.
Common Core State Standards Initiative. (2010). *Mathematics standards*. Retrieved on March 07, 2021 from http://www.corestandards.org/Math/.
Cornbleth, C. (1996). Curriculum in and out of context. In E. R. Hollins (Ed.), *Transforming curriculum for a culturally diverse society* (pp.149–161). Mahwah, NJ: Lawrence Erlbaum.
Darling-Hammond, L. (1990). Instructional policy into practice: "The power of the bottom over the top." *Educational Evaluation and Policy Analysis, 12*(3), 339–347.
Drake, C. (2002). Experience counts: Career stage and teachers' responses to mathematics education reform. *Educational Policy, 16*(2), 311–337.
Denzin, N. K., & Lincoln, Y. S. (2005). Introduction: The discipline and practice of qualitative research. In N. K. Denzin & Y. S. Lincoln (Eds.), *The SAGE handbook of qualitative research* (pp.1–32). Thousand Oaks, CA: Sage Publications.
Desimone, L., Smith, T., & Phillips, K. (2007). Does policy influence mathematics and science teachers' participation in professional development? *Teachers College Record, 109*(5), 1086–1122.
Doyle, W., & Ponder, G. (1977–1978). The practicality ethic in teacher decision-making. *Interchange, 8*(3), 1–12.
Duncan, A. (2009, February 9). *Secretary Arne Duncan speaks at the 91st annual meeting of the American Council on Education*. Retrieved on October 10, 2020 from www.ed.gov/news/speeches/secretary-arne-duncan-speaks-91st-annual-meeting-american-council-education.
Fan, L., Wong, N.-Y., Cai, J. & Li, S. (Eds.). (2004). *How Chinese learn mathematics: Perspectives from insiders*. Singapore: World Scientific Publishers.
Fosnot, C. T. (2005). Constructivism revisited: Implications and reflections. *The constructivist, 16*(1).
Frankenstein, M. (1992). Critical mathematics education: An application of Paulo Freire's epistemology. In K. Weiler & C. Mitchell (Eds.), *What schools can do: Critical pedagogy and practice* (pp.237–264). Albany, NY: State University of New York Press.
Fuhrman, S. H. (1999). *The new accountability. CPRE Policy Briefs*. Retrieved on March 7, 2021 from http://repository.upenn.edu/cpre_policybriefs/73.

Fullan, M. G. (1991). *The new meaning of educational change.* New York, NY: Teachers College Press.

Fullan, M., & Pomfret, A. (1977). Research on curriculum and instruction implementation. *Review of Educational Research, 47*(2), 335–397.

Ginsburg, M. B., Cooper, S., Raghu, R., & Zegarra, H. (1990). National and world-system explanations of educational reform. *Comparative Education Review, 34* (4), 474–499.

Guan, Q. & Meng, W. J. (2007). China's new national curriculum reform: Innovation, challenges and strategies. *Frontiers of Education in China, 2*, 579–604.

Guskey, T. R. (1999). Moving from means to ends. *Journal of Staff Development, 20*(2), 48.

Guskey, T. R. (2002). Professional development and teacher change. *Teachers and Teaching, 8*(3), 381–391.

He, X. Y. (2006). Response to "Jiang Boju: What is lost in mathematics class due to the new Standards?" *Guang Dong Education*, p.6.

He, X. Y. (2008). The Educational War and the road of mathematics education. *Journal of Mathematics Education, 17*(1).

Huang, F. Q. (2004). Curriculum reform in contemporary China: Seven goals and six strategies. *Journal of Curriculum Studies, 36*(1), 101–115.

Huberman, A. M., & Miles, M. B. (1983). Drawing valid meaning from qualitative data: Some techniques of data reduction and display. *Quality and Quantity, 17*, 281–339.

Ingersoll, R. M. (2011). Power, accountability, and the teacher quality problem. In S. Kelly (Ed.), *Assessing teacher quality: Understanding teacher effects on instruction and achievement* (pp.97–109). New York, NY: Teachers College Press.

Jiang, B. J. (2005, March 16). What is lost in mathematics class due to the new Standards? *Guangming Daily-Educational Weekly,* p.5.

Johnson, B., & Turner, L. A. (2003). Data collection strategies in mixed methods research. In A. Tashakkori & C. Teddlie (Eds.), *SAGE handbook of mixed methods in social & behavioral research* (pp.297–319). Thousand Oaks, CA: Sage Publications.

Liu, D. J. (2007, April 18). *Why is the burden of children getting more and more heavy? Xinhua Journal*, p.1.

Liu, S., & Teddlie, C. (2005). A follow-up study on teacher evaluation in China: Historical analysis and latest trends. *Journal of Personnel Evaluation in Education, 18*, 253–272.

Lortie, D. C. (1975). *Schoolteacher: A sociological study*. Chicago, IL: University of Chicago Press.

Lu, J., & Wang, Y. (2004). The new edition of Chinese mathematics textbooks for primary schools. Tenth International Congress on Mathematical Education (ICME-10), Copenhagen, Denmark.

Ma, L. P. (1999). *Knowing and teaching elementary mathematics: teachers' understanding of fundamental mathematics in China and the United States*. Mahwah, NJ: Lawrence Erlbaum.

Manouchehri, A. (1997). School mathematics reform: Implications for mathematics teacher preparation. *Journal of Teacher Education, 48*(3), 197–209.

Mason, B., Mason, D. A., Mendez, M., Nelsen, G., & Orwig, R. (2005). Effects of top-down and bottom-up elementary school standards reform in an underperforming California district. *The Elementary School Journal, 105*(4), 353–77.

Merits School. (1984). *Reports on reform experience of learning Ma Xinlan*. Internal report: Unpublished.

Merits School. (1993). *The annual plans and summaries.* Internal report: Unpublished.

Merits School. (2000). *The annual plans and summaries.* Internal report: Unpublished.

Merits School. (2002). *The teaching evaluation system.* Internal report: Unpublished.

Merits School. (2003). *The 2002–2003 school report.* Internal report: Unpublished.

Merits School. (2004). *The self-evaluation report to the D.E.B.* Internal report: Unpublished.

Merits School. (2009a). *Evaluation results of teachers' lesson plans.* Internal report: Unpublished.

Merits School. (2009b, April 20). *The evaluation records and summaries of Grade 1–5 lesson plans, student work, and lesson observation notes.* Internal report: Unpublished.

Merriam, S. B. (1998). *Qualitative research and case study applications in education.* San Francisco, CA: Jossey-Bass Publishers.

Meyer, H-D., & Rowan, B. (Eds.) (2006). *The new institutionalism in education.* Albany, NY: State University of New York Press.

M.O.E. (1999, June 13). *The decision to strengthen the educational reform and to foster full-scale qualities-oriented education.* Retrieved on March 10, 2020 from www.moe.gov.cn/jyb_sjzl/moe_177/tnull_2478.html.

M.O.E. (2001a). *Curriculum standards for school mathematics of nine-year compulsory education (trial version).* Beijing: Beijing Normal University Press.

M.O.E. (2001b). *The curriculum reform outline for basic education (trial).* Retrieved on October 16, 2020 from http://old.moe.gov.cn/publicfiles/business/htmlfiles/moe/moe_309/200412/4672.html.

M.O.E. (2001c). *The trial Mathematics Curriculum Standards for full-time compulsory education.* Beijing: Beijing Normal University Press.

M.O.E. (2002). *The notice about actively promoting the reform of the system of assessment and examination in elementary and middle schools.* Retrieved on March 12, 2021 from http://www.moe.gov.cn/srcsite/A26/s7054/200212/t20021218_78509.html.

M.O.E. (2007). *The revised Mathematics Curriculum Standards for full-time compulsory education.* Beijing: People's Education Press.

M.O.E. (2014). *Opinions on deepening the curriculum reform to fulfil the fundamental task of cultivating moral and competent persons through education.* Retrieved on August 10, 2020 from http://old.moe.gov.cn/publicfiles/business/htmlfiles/moe/s7054/201404/167226.html.

National Council of Teachers of Mathematics. (1989). *Curriculum and evaluation standards for school mathematics.* Reston, VA: National Council of Teachers of Mathematics.

National Research Council. (1990). Reshaping *s*chool *m*athematics: A *p*hilosophy and *f*ramework for *c*urriculum. Washington, DC: The National Academies Press.

O'Brien, T. C. (1999). Parrot math. *Phi Delta Kappan, 80*(6), 434–438.

Ogawa, R. T., Sandholtz, J. H., Martinez-Flores, M., & Scribner, S. P. (2003). The substantive and symbolic consequences of a district's standards-based curriculum. *American Educational Research Journal, 40*(1), 147–176.

Pajares, M. F. (1992). Teachers' beliefs and educational research: cleaning up a messy construct. *Review of Educational Research, 62*(3), 307–332.

Parise, L. M., & Spillane, J. P. (2010). Teacher learning and instructional change: How formal and on-the-job learning opportunities predict change in elementary school teachers' practice. *The Elementary School Journal, 110*(3), 323–346.

Penuel, W. R., Fishman, B. J., Yamaguchi R. Y., & Gallagher, L. P. (2007). What Makes Professional Development Effective? Strategies that Foster Curriculum Implementation. *American Education Research Journal, 44*(4), 921–958.

Peressini, D. D. (1998). The portrayal of parents in the school mathematics reform literature: Locating the context for parental involvement. *Journal for Research in Mathematics Education, 29*(5), 555–582.

Pioneer School. (1999). *The academic assessment system for elementary mathematics education*. Internal report: Unpublished.

Pioneer School. (2003a). *The collection of documents regarding the I.E.M.E. Experiment*. Internal report: Unpublished.

Pioneer School. (2003b). *Regulations and measures on teaching evaluation*. Internal report: Unpublished.

Pioneer School. (2003c). *The records of Grade 2 teaching research month activities*. Internal report: Unpublished.

Pioneer School. (2004). *The records of school-level teaching research activities*. Internal report: Unpublished.

Pioneer School. (2008). *The records of school-level teaching research activities*. Internal report: Unpublished.

Porter, A. C. (1994). National standards and school improvement in the 1990s: Issues and promise. *American Journal of Education, 102*(4), 421–449.

Prediger, S. (2004). Intercultural perspectives on mathematics learning – developing a theoretical framework. *International Journal of Science and Mathematics Education, 2*, 377–406.

Price, J., Ball, D., & Luks, S. (1995). *Marshaling resources for reform: District administrators and the case of mathematics*. East Lansing, MI: National Center for Research on Teacher Learning.

Province Teacher Training Center. (n.d.). *Notice on selecting teachers to attend the training on the new curriculum reform*. Internal report: Unpublished.

Red Pebble D.E.B. (2002). *Fully implementing the new curriculum reform*. Internal report: Unpublished.

Red Pebble District T.R. Center. (2002). *The comprehensive assessment plan for elementary schools*. Internal report: Unpublished.

Remillard, J. T., & Jackson, K. (2006). Old math, new math: Parents' experiences with Standards-based reform. *Mathematical Thinking and Learning, 8*(3), 231–259.

Sargent, T. C., & Hannum, E. (2009). Doing more with less: Teacher professional learning communities in resource-constrained primary schools in rural China. *Journal of Teacher Education, 60*(3), 258–276.

Sargent, T. C. (2011). New curriculum reform implementation and the transformation of educational beliefs, practices, and structures in Gansu Province. *Chinese Education and Society, 44*(6), 47–72.

Schoenfeld, A. H. (2004). The math wars. *Educational Policy, 18*(1), 253–286.

Smith, M. S., & O'Day, J. (1991). Systemic school reform. In S. H. Fuhrman, & B. Malen (Eds.), *The politics of curriculum and testing* (pp.233–267). London: Taylor & Francis.

Spillane, J. P. (1996). School districts matter: Local educational authorities and state instructional policy. *Educational Policy, 10*(1), 63–87.

Spillane, J. P. (1999). External reform initiatives and teachers' efforts to reconstruct their practice: The mediating role of teachers' zones of enactment. *Journal of Curriculum Studies, 31*(2), 143–175.

Spillane, J. P., Hallett, T. & Diamond, J. (2003). Forms of capital and the construction of leadership: Instructional leadership in urban elementary schools. *Sociology of Education, 76*(1), 1–17.

Spillane, J. P., Reiser, B., & Gomez, L. M. (2006). Policy implementation and cognition: The role of human, social, and distributed cognition in framing policy implementation. In M. I. Honig (Ed.), *New directions in educational policy implementation* (pp.47–64). Albany, NY: State University of New York Press.

Spillane, J. P. & Thompson, C. L. (1997). Reconstructing conceptions of local capacity: The local education agency's capacity for ambitious instructional reform. *Educational Evaluation and Policy Analysis, 19*(2), 185–203.

Spillane, J. P. & Zeuli, J. S. (1999). Reform and teaching: Exploring patterns of practice in the context of national and state mathematics reforms. *Educational Evaluation and Policy Analysis, 21*(1), 1–27.

Stake, R. E. (1995). *The art of case study research.* Thousand Oaks, CA: Sage Publications.

Steffe, L., Kieren, T. (1994). Radical constructivism and mathematics education. *Journal for Research in Mathematics Education, 25*(6), 711–733.

Stenhouse, L. (1975). *An introduction to curriculum research and development.* London: Heinemann.

Stevenson, H. W. & Stigler, J. W. (1994). *The learning gap: Why our schools are failing and what we can learn from Japanese and Chinese education.* New York, NY: Simon & Schuster.

Stigler, J. W. & Hiebert, J. (1999). *The teaching gap: Best ideas from the world's teachers for improving education in the classroom.* New York, NY: Free Press.

Stinson, D. W. & Bullock, E. C. (2012). Critical postmodern theory in mathematics education research: A praxis of uncertainty. *Educational Studies in Mathematics, 80*(1–2), 41–55.

Stryker, S. (2008). From Mead to a structural symbolic interactionism and beyond. *Annual Review of Sociology, 34*, 15–31.

Sun, L. K. (1991). Contemporary Chinese culture: Structure and emotionality. *The Australian Journal of Chinese Affairs, 26*, 1–41.

Sun, X. T. (2005). Several Whys on the New Mathematics Curriculum—The interview of Professor Sun Xiaotian by Yu Huijuan. *People's Education,* (7).

Tsang, M. C. (2000). Education and national development in China since 1949: Oscillating policies and enduring dilemmas. *China Review,* 579–618.

Van de Ven, A. H. (2007). *Engaged scholarship: A guide for organizational and social research.* New York, NY: Oxford University Press.

Wang, J. & Lin, E. (2005). Comparative studies on U.S. and Chinese mathematics learning and implications for standards-based mathematics teaching reform. *Educational Researcher, 34*(5), 3–13.

Whyte, W. F. (1993). *Street corner society: The social structures of an Italian slum.* Chicago, IL: University of Chicago Press.

Yin, R. K. (2003). *Case study research and applications: Design and methods (3rd Ed.).* Thousand Oaks, CA: Sage Publications.

Yoshida, M. (1999). *Lesson study: An ethnographic investigation of school-based teacher development in Japan* [Unpublished doctoral dissertation]. University of Chicago.

Yuan, G. R. (2005). *Providing strong teacher support for the New Curriculum Reform.*

Zhong, Q. Q. (2006, December 15). Determined and unyielding: Trumpeting the Reform March. *China Education Daily*, p.5.

Index

Note: Page numbers in *italics* indicate figures and in **bold** indicate tables on the corresponding pages.

For Product Safety Concerns and Information please contact our EU representative GPSR@taylorandfrancis.com
Taylor & Francis Verlag GmbH, Kaufingerstraße 24, 80331 München, Germany

www.ingramcontent.com/pod-product-compliance
Lightning Source LLC
LaVergne TN
LVHW021127300926
849160LV00010B/893
* 9 7 8 1 0 3 2 0 2 3 3 6 6 *